Self-Commissioning Wireless Power Transfer

This accessible guide describes the design and implementation of new front-end monitoring and control strategies for wireless power transfer (WPT). The new control paradigm moves all receiver-side controllers to the transmitter side to enable multiple advantages over conventional designs.

Self-Commissioning Wireless Power Transfer: A New Control Paradigm provides readers with information in three areas. First, it discusses a series of front-end monitoring and control strategies for WPT. The new technologies are designed at the transmitter side, while no control is needed at the receiver side. In addition, no communication is required between the transmitter and the receiver. Implementing this change will enable engineers to make the receiver more compact, reliable, and cost-effective and enhance the overall efficiency of the system as the main power circuits of the WPT system, auxiliary power supplies of controllers, and protection circuits can be significantly reduced. Second, it presents designs for new control strategies in achieving maximum efficiency transfer and minimum charging time, along with precise modeling of battery loads and other power electronic components throughout the charging process. Third, it illustrates how new control methodologies integrate with other technologies for compatibility in industrial applications. Throughout, this book illustrates its descriptions of the technologies and designs with real-world case studies and analysis.

Researchers working on WPT technologies in academia and industry will benefit immensely from the discussions presented in this book. Graduate students aiming to increase their knowledge on control schemes for power converters at the transmitter side and primary-side control for transmitter-side inverters will also find it valuable.

Yun Yang is an assistant professor at Nanyang Technological University, Singapore. He has published over 100 technical papers, including more than 50 in top-tier journals on wireless power transfer, many of which have been recognized as best transaction or conference papers.

Self-Commissioning Wireless Power Transfer

A New Control Paradigm

Yun Yang

CRC Press
Taylor & Francis Group
Boca Raton London New York

CRC Press is an imprint of the
Taylor & Francis Group, an **informa** business

Designed cover image: alengo

First edition published 2026
by CRC Press
2385 NW Executive Center Drive, Suite 320, Boca Raton FL 33431

and by CRC Press
4 Park Square, Milton Park, Abingdon, Oxon, OX14 4RN

CRC Press is an imprint of Taylor & Francis Group, LLC

ISBN: 9781032737812 (hbk)
ISBN: 9781032742823 (pbk)
ISBN: 9781003468523 (ebk)

DOI: 10.1201/9781003468523

Typeset in Palatino
by codeMantra

For my daughter,

Yeuk Ham

Contents

1

Control Paradigm Shift for Next-Generation Inductive-Type Wireless Power Transfer

1.1 Evolution of Inductive Wireless Power Transfer

The Tesla coil, invented by Nikola Tesla in 1891, is considered the earliest known inductive-type wireless power transfer (WPT) system [1]. Tesla demonstrated wireless power transmission through the air using the principles of magnetic resonance, high-frequency operation, low winding resistance, and high quality (Q) factors [2]. These design principles for the inductance-capacitance (LC) resonant circuit in the Tesla coil remain relevant in today's WPT systems. After Tesla's pioneering work, research in WPT remained relatively subdued until significant advancements in the 1970s with the introduction of Litz wires and in the 1980s with the development of semiconductor switches. Initially, WPT applications focused on biomedical device charging through tissues, exemplified by the invention of the pacemaker by Prof. Bo Cho in the late 1980s [3]. By the 1990s, WPT technologies began expanding into various applications including electric vehicles (EVs), mobile robots, and consumer electronics. Pioneers in the field such as Prof. John Boys, Prof. Ron Hui, and Prof. Chun T. Rim made notable contributions: Prof. Boys in inductive WPT for mobile robots [4], Prof. Hui in planar coreless transformers [5], and Prof. Rim in road-powered EV systems [6]. Figure 1.1 illustrates the prototype of the planar wireless charging system with free positioning developed by Prof. Hui's research team.

The advent of mobile phones in the 1990s accelerated the evolution of WPT [7]. Concepts for WPT in portable electronics, utilizing magnetically coupled windings, resonance techniques, and power electronics, were first documented in [8–14]. Among the various design approaches explored, planar wireless charging platforms using vertical flux [15] and horizontal flux [16] were particularly prominent in the early 2000s. The horizontal flux approach employs a compacted winding structure resembling a rational machine, where magnetic flux travels horizontally across the charging pad surface. Consequently, the receiver coil must align vertically with the pad to capture this flux, posing challenges for slim designs in portable electronics. In contrast, the vertical flux approach directs magnetic flux vertically from the charging pad surface. This allows for flexible placement of the receiver coil, making it more suitable for slim mobile phone designs. This advantage led to

DOI: 10.1201/9781003468523-1

FIGURE 1.1
Photograph of the planar wireless charger developed by Prof. Hui's research team for Philips. (Reproduced with permission from Prof. Ron Hui.)

the adoption of the vertical flux approach by the world's first wireless charging standard "Qi" introduced by the Wireless Power Consortium (WPC) in 2010 [17]. To date, over 8500 products have received "Qi" certification, underscoring its widespread adoption and impact on the field of WPT. The introduction of the "Qi" standard also sparked intensive research activities in WPT, driving further advancements in the technology. Figure 1.2 shows the proliferation of journal publications on inductive-type WPT. Evidently, the number of publications saw a sharp increase after 2010. However, I posit that the 2-meter WPT experiment conducted by the Massachusetts Institute of Technology (MIT) research team in 2007, which preceded this surge by three years, also significantly catalyzed investigations into magnetic resonance-based WPT [18].

Subsequently, two alternative WPT standards were introduced by the AirFuel Alliance (AFA) and the Society of Automotive Engineers (SAE), known respectively as "A4WP" and "TIR J2954", targeting high-frequency electronic devices [19] and high-power EVs [20]. However, due to technological immaturity, adoption of these standards has been limited, unlike the widespread adoption of the "Qi" standard established by the WPC. The "Qi" standard remains predominant. In recent years, the WPC has expanded

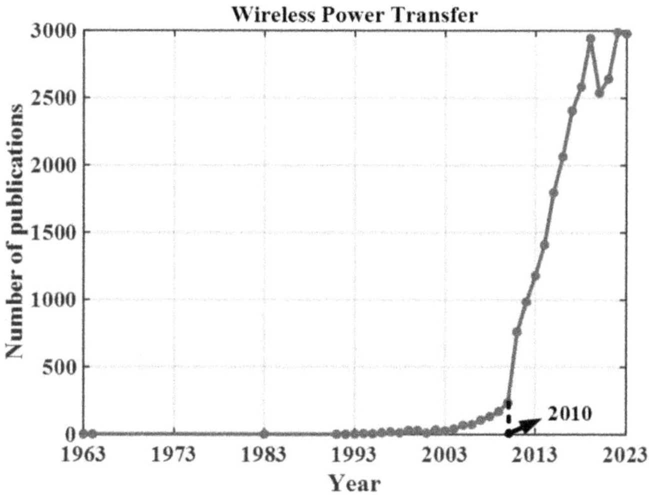

FIGURE 1.2
Number of journal publications on inductive-type WPT since 1963.

its scope to include higher power applications, such as the "Ki" standard for cordless kitchen devices [21] and the "LEV" standard for light EVs [22]. However, these standards, while ensuring "compatibility", often prioritize "compatibility" over "optimal performance".

1.2 Next-Generation WPT

"Optimal performance" represents a broad concept defining the primary goals of next-generation WPT, clearly distinguishing it from the "compatibility" focus of current WPT standards. In this context, "optimal performance" in this book specifically denotes achieving (i) maximum efficiency transfer (MET) and/or (ii) minimum charging time (MCT). The rationale for targeting MET and MCT in next-generation WPT systems stems from the significant advancements in power levels and frequencies enabled by WBG semiconductors. Even marginal improvements in efficiency—such as a 1% increase—can lead to substantial reductions in carbon footprint. Consequently, minimizing energy wastage has become a fundamental product requirement in many regions striving for carbon neutrality. Recently, the WPC introduced the "Qi2" standard, which received widespread acclaim at the Consumer Electronics Show in 2023. The "Qi2" standard prioritizes MET and MCT as critical factors. Figure 1.3 illustrates the power-frequency domains of current WPT applications covered by the WPC, SAE, and AFA. It is evident that the existing "Qi", "SAE J2954", and "AirFuel" standards operate within established frequency and power ranges,

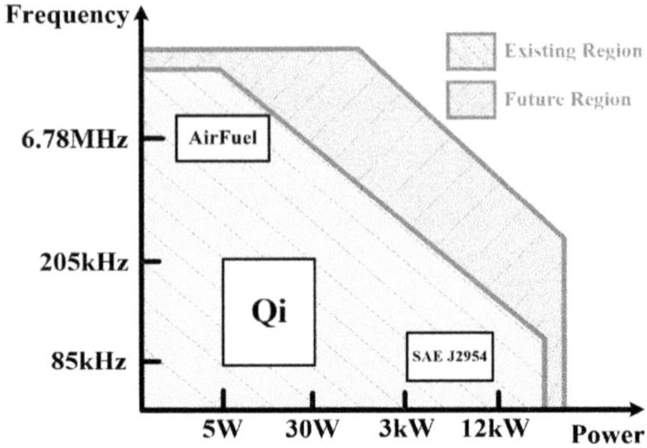

FIGURE 1.3
Power-frequency regions of current and future WPT applications.

leaving a significant unexplored domain with higher power and frequencies. The future evolution of the "Qi2" standard aims to extend coverage into these uncharted territories. Figure 1.3 illustrates the power-frequency domains currently addressed by WPT applications under the auspices of the WPC, SAE, and AFA. It is evident that the current standards—such as "Qi", "SAE J2954", and "AirFuel"—are confined within established operational ranges. However, there exists a substantial unexplored domain characterized by higher power levels and frequencies. Future iterations of standards, like "Qi2", have the potential to expand into these higher power and frequency regions, thereby extending the capabilities of WPT technologies [23].

1.3 Basic Principle of MET

It is widely recognized that Q factors and coupling coefficients are pivotal in determining the MET of a typical WPT system. Higher Q factors and stronger coupling coefficients are initially designed to achieve higher peak efficiencies. However, in practical operation, achieving MET in a WPT system involves optimizing the equivalent load impedance to an optimal value.

To illustrate this concept without losing generality, consider a two-stage series-series (SS)-compensated WPT system. Figure 1.4 shows the equivalent M-model of the resonators with fundamental frequency components. Here, v_p represents the input AC voltage, i_p and i_s denote the transmitter and receiver currents respectively, L_p and L_s are the self-inductances of the transmitter and receiver coils, and C_p and C_s are the compensated capacitances designed to nullify the reactance at the resonant frequency $\omega_o = \frac{1}{\sqrt{L_p C_p}} = \frac{1}{\sqrt{L_s C_s}}$.

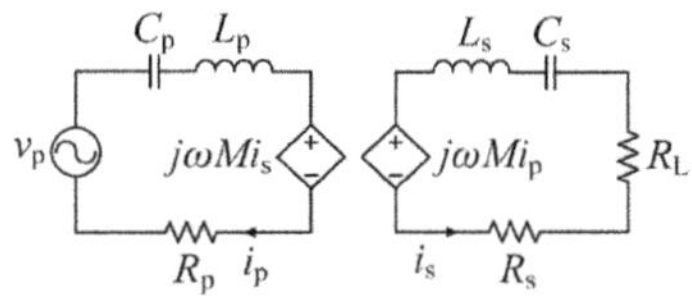

FIGURE 1.4
Equivalent M-model of the resonators with fundamental frequency components.

Additionally, R_p and R_s represent the equivalent series resistances (ESRs) of the transmitter and receiver resonators, while R_L stands for the equivalent load resistance that can be regulated by a load-end power converter. Using this equivalent circuit, the system's efficiency can be derived as follows:

$$\eta = \frac{I_s^2 R_L}{I_s^2 (R_L + R_s) + I_p^2 R_p} \tag{1.1}$$

where I_p and I_s represent the root-mean-square (RMS) values of i_p and i_s, respectively. The relationship between the transmitter and receiver currents at the resonant frequency can be established from the right-hand side of the equivalent circuit as

$$j\omega M i_p + (R_s + R_L) i_s = 0 \tag{1.2}$$

By substituting Equation (1.2) into Equation (1.1) to eliminate I_p,

$$\eta = \frac{\dfrac{\omega^2 M^2}{R_p}}{R_L + \dfrac{\left(R_s + \dfrac{\omega^2 M^2}{R_p}\right) R_s}{R_L} + 2R_s + \dfrac{\omega^2 M^2}{R_p}} \tag{1.3}$$

Then, the optimal load resistance for achieving MET can be determined from Equation (1.3) as

$$R_{LMET} = R_s \sqrt{1 + \frac{\omega^2 M^2}{R_p R_s}} \tag{1.4}$$

The corresponding MET is

$$\eta_{MET} = \frac{\dfrac{\omega^2 M^2}{R_p R_s}}{\left(1 + \sqrt{1 + \dfrac{\omega^2 M^2}{R_p R_s}}\right)^2} \tag{1.5}$$

It is important to distinguish between MET and maximum power transfer (MPT). Using the same circuit depicted in Figure 1.4, we can illustrate this difference. According to the equivalent circuit, the output power of the system is

$$P_o = I_s^2 R_L \tag{1.6}$$

According to Kirchhoff's laws, at the resonant frequency,

$$v_p = j\omega M i_s + R_p i_p \tag{1.7}$$

By substituting Equation (1.2) and Equation (1.7) into Equation (1.6),

$$P_o = \frac{\omega^2 M^2 V_p^2}{R_p^2 R_L + \frac{\left(\omega^2 M^2 + R_p R_s\right)^2}{R_L} + 2R_p\left(\omega^2 M^2 + R_p R_s\right)} \tag{1.8}$$

Then, the optimal load resistance for achieving MPT can be determined from Equation (1.8) as

$$R_{\mathrm{LMPT}} = R_s\left(1 + \frac{\omega^2 M^2}{R_p R_s}\right) \tag{1.9}$$

The corresponding MPT is

$$P_{\mathrm{oMPT}} = \frac{V_p^2}{4R_p\left(1 + \frac{R_p R_s}{\omega^2 M^2}\right)} \tag{1.10}$$

Clearly, the optimal load condition for MPT in Equation (1.9) differs from that for MET in Equation (1.4). By substituting Equation (1.9) into Equation (1.3), the corresponding system efficiency can be calculated by

$$\eta_{\mathrm{MPT}} = \frac{\omega^2 M^2}{2\omega^2 M^2 + 4R_p R_s} \tag{1.11}$$

Obviously, the maximum value of η_{MPT} is 50% (i.e., $R_p = R_s = 0\ \Omega$).

In a highly cited paper, Hui et al. [24] advocate for using MET rather than MPT in near-field WPT applications. This recommendation stems from the inherent limitation of MPT, where at least half of the input energy is dissipated in the source resistance. It's important to note that this 50% efficiency constraint differs from the maximum value of η_{MPT} described in Equation (1.11). The 50% efficiency cited in Hui et al. [24] pertains to WPT systems powered by radio frequency (RF) power amplifiers, where achieving 50 Ω

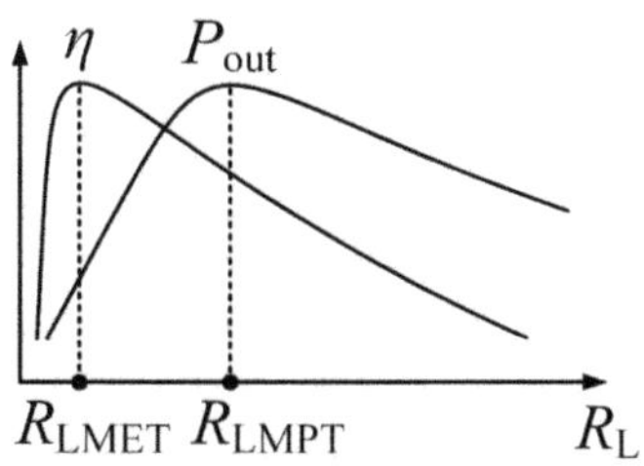

FIGURE 1.5
Efficiency-load and output power-load characteristics of an SS-compensated WPT system.

impedance matching is typically necessary. Therefore, the 50% efficiency limit is imposed by this impedance requirement. On the other hand, the maximum value of η_{MPT} being 50% indicates the highest efficiency achievable by the coupled resonator when the output power of the system is optimized at the optimal load condition. Figure 1.5 displays the efficiency and output power curves of a typical two-stage SS-compensated WPT system under various load conditions. The equivalent load resistance R_L is recommended to be adjusted within the ranges of R_{LMET} and R_{LMPT}. The choice of R_L depends on the primary control objective. To prioritize energy conservation, R_L can be set at R_{LMET}. Conversely, for rapid charging, R_L can be set to R_{LMPT}.

1.4 Basic Principle of MCT

Charging time is a direct reflection of charging power. However, equating MCT to maximum output power (MOP) is inaccurate due to various constraints such as power, current, voltage, magnetic field, and temperature limits that may prevent reaching the MOP. Temperature constraints, particularly for battery loads, must be rigorously managed to ensure safe operation. The Japan Electronics and Information Technology Industries Association (JEITA) has established an early battery temperature protocol, illustrated by a typical charging profile of a lithium-ion (Li-ion) battery compliant with JEITA guidelines shown in Figure 1.6a. According to this guideline, charging power is halved when the battery temperature reaches its limit, prioritizing safety over charging speed. Consequently, the average charging power falls well below the MOP, resulting in longer than expected charging times. To mitigate this issue without compromising safety, a continuous temperature control law was proposed in [25]. This strategy aims to reduce average charging currents while keeping RMS values unchanged. Figure 1.6b depicts a charging profile based on this approach, where battery temperature closely follows a fixed reference, enabling faster charging compared to the JEITA guideline. However, this continuous temperature

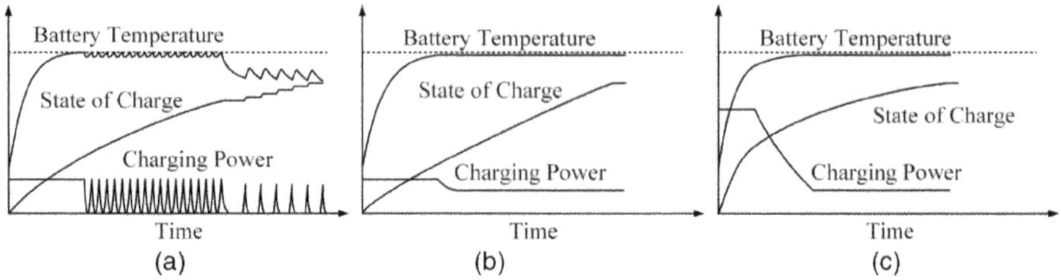

FIGURE 1.6
Typical charging profiles based on (a) JEITA guideline, (b) the continuous temperature control law, and (c) Qualcomm Quick Charge.

control law overlooks the initial temperature inertia, where a conservative constant charging power is initially applied until the temperature stabilizes. In practice, the charging power may start at five to six times the nominal value before stepping down to ensure the battery temperature remains below the upper limit (as seen in Qualcomm Quick Charge standards). This dynamic adjustment further reduces total charging time, as demonstrated in Figure 1.6c.

In summary, the fundamental principle to achieve MCT is to prioritize the safety of WPT operations, including stringent control of battery temperatures within safe limits, followed by maximizing the average output power.

1.5 Existing Control Paradigm

The current methods for achieving MET in WPT systems can be categorized into two main approaches: direct MET control and indirect MET control. Direct MET control methods "directly" adjust the equivalent load impedance of the load-end DC-DC converters and actual loads to optimal values (i.e., R_{LMET}) using the duty ratios of the converters [26–28]. The duty ratio control can be implemented with feedback loops from both transmitters and receivers, or solely from receivers, as illustrated in Figure 1.7. When feedback is solely from receivers, phase shift or frequency control strategies are typically employed in transmitter-side driving circuits, with or without direct communication between transmitters and receivers.

The dual feedback control method depicted in Figure 1.7a operates by evaluating system efficiency through measured input and output power (i.e., P_{in} and P_{out}). Given that system efficiency varies monotonically with the equivalent load resistance, the optimal duty ratio of the DC-DC converter can be determined using a search algorithm such as the perturb and observe (P&O) method. This control approach offers robustness against parameter drifts without requiring complex equations. The control signal directly correlates

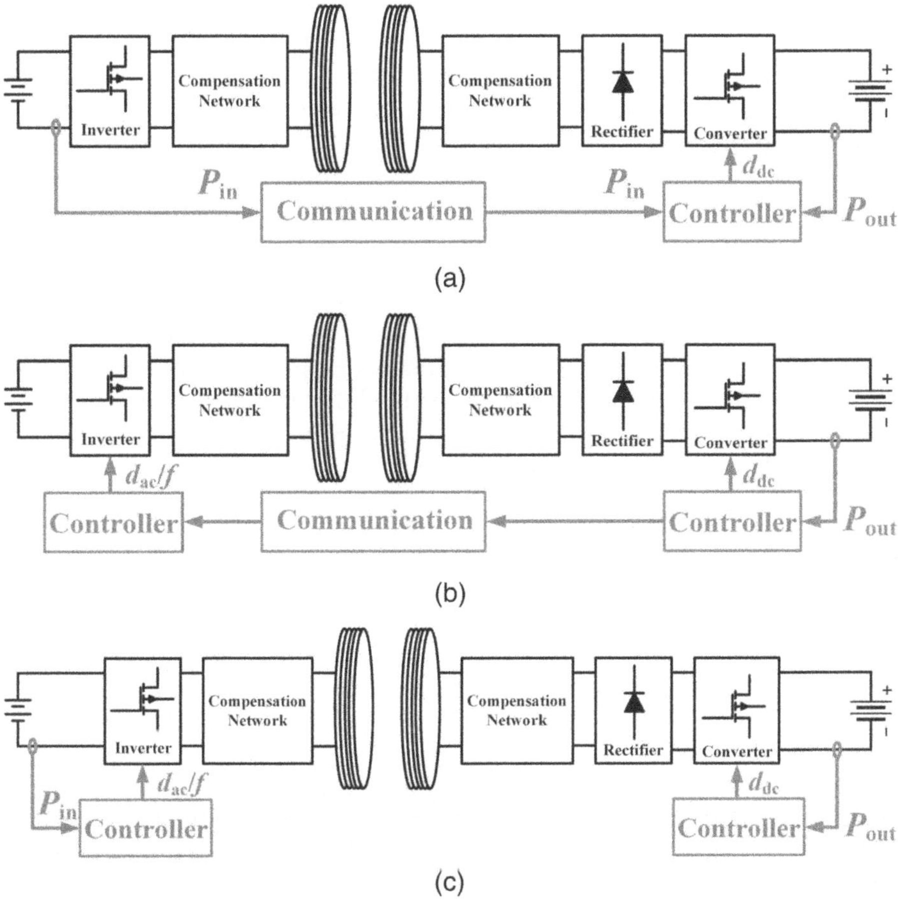

FIGURE 1.7
Schematic diagrams of direct MET control (a) with transmitter-to-receiver communication, (b) with receiver-to-transmitter communication, and (c) without communication between the transmitter and the receiver.

with the control objective. However, a notable drawback is its reliance on an out-of-band communication channel, which can transmit inaccurate information about the input power P_{in}.

Similar challenges are also present in the communication-based approach depicted in Figure 1.7b. Here, the DC-DC converter adjusts the output voltage, while the power inverter on the transmitter side perturbs the operating point by altering the conversion ratio d_{ac} or frequency f to track the maximum efficiency point. An advantage of this method over the approach in Figure 1.7a is the consideration of efficiency optimization on the transmitter-side inverter. Soft switching of the inverter can be achieved through the regulation of either d_{ac} or f.

The communication-free method illustrated in Figure 1.7c shares its foundation with the approach shown in Figure 1.7b. The core control principle for the load-end DC-DC converter aligns with methods depicted in Figure 1.7a and b, focusing on optimizing efficiency. However, the methodology diverges in how the transmitter-side inverter is regulated. In the communication-free control method, minimizing P_{in} is achieved through a P&O algorithm [29] or utilizing in-band information transmitted from the receiver side [30]. A drawback of the communication-free approach in Figure 1.7c, compared to the communication-based method in Figure 1.7b, is its slower dynamic response due to narrower bandwidth constraints.

Indirect MET control methods involve regulating system efficiency indirectly by adjusting other electrical parameters close to their optimal values. For instance, in the WPT system described in [31], MET can be approximated by adjusting the voltage ratio between the transmitter and receiver resonators to satisfy

$$\frac{|v_s|}{|v_p|} = \sqrt{\frac{R_s}{R_p}} \tag{1.12}$$

This approach is implemented slightly differently in [32], where the phase angle between v_p and v_s is controlled. Another method utilizing currents to achieve MET is demonstrated in [33], where the phase angle between the transmitter current (i.e., i_p) and receiver voltage (i.e., v_s) is regulated based on a phase-locked loop.

Table 1.1 presents a comparative analysis of typical MET control methods reported over the last decade. These prominent methods exhibit key features such as (i) optimizing efficiency for entire systems, (ii) maintaining robustness against parameter drifts and disturbances, (iii) operating without wireless communication, (iv) enabling fast dynamics, and (v) implementing soft switching for all power switches. Despite these strengths, none of these methods fully meet the requirements for receivers that are compact, reliable, and cost-effective. For high-power and higher-frequency WPT systems, we propose that a compact, reliable, and cost-effective receiver should consist solely of a passive rectifier, a filter, and basic protection circuits. The new control paradigm for the next-generation WPT systems should encompass all the features listed in Table 1.1 to address the evolving demands of efficiency, reliability, and simplicity in receiver design.

Existing MCT control methods for WPT systems are typically structured around a cascaded temperature-power architecture, illustrated in Figure 1.8. In this setup, outer-loop temperature control sets the reference for inner-loop power control based on measured or estimated battery temperature T_b. The location of sensors and application specifics determine whether T_b represents surface temperature or internal temperature. The temperature controller can adopt a model-free or model-based approach. Model-free control, such as classic proportional-integral (PI) control [25], is commonly used to maintain

TABLE 1.1

Comparison of Typical MET Control Methods

Feature	**Ref. [26]**	**Ref. [27]**	**Ref. [28, 30]**	**Ref. [29]**	**Refs. [31–33]**	**Ref. [34]**	**Ref. [35]**	**Ref. [36]**	**Ref. [37]**	**Ref. [38]**
Efficiency optimization for entire systems	×	√	√	√	×	×	×	√	√	√
Robustness against parameter drifts and disturbance	√	√	√	√	×	√	√	×	×	×
No wireless communication	×	×	√	√	×	√	×	√	×	×
Fast dynamics	√	√	√	×	√	√	√	×	×	√
Soft switching for all power switches	×	√	×	×	×	×	√	√	√	√
Compact, reliable, and cost-effective receivers	×	×	×	×	×	×	×	×	×	×

Source: Modified from reference [41], with permission.

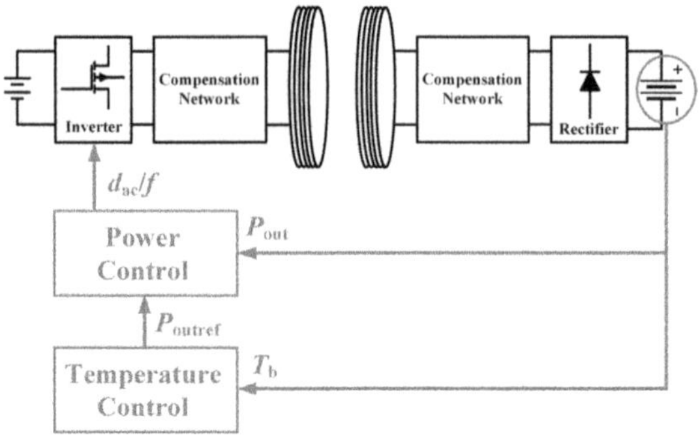

FIGURE 1.8
Schematic diagram of a typical MCT control.

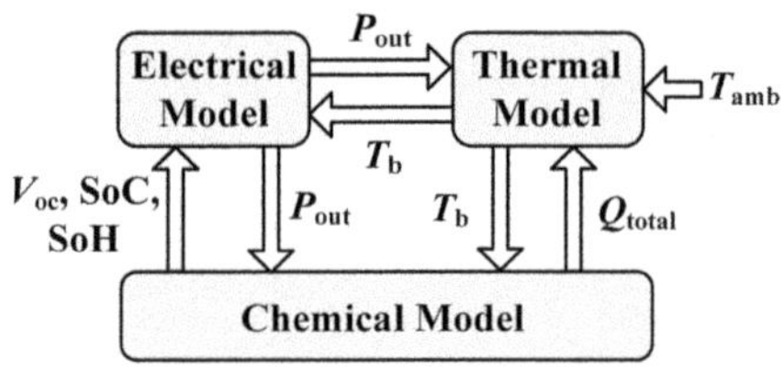

FIGURE 1.9
Schematic diagram of a conceptual electrical-thermal-chemical model.

battery temperature within specified limits. Model-based temperature control utilizes linear or nonlinear strategies with support from mapping functions [39], artificial intelligence (AI)-based models, or circuit-emulated electrothermal models [40]. Electrothermal models can be enhanced by integrating chemical modules or AI-based adjustments for improved performance. Figure 1.9 illustrates a conceptual electro-thermal-chemical model for Li-ion batteries, where Q_{total} represents total heat generation comprising irreversible thermal dissipation and reversible chemical reaction heat. T_{amb} denotes ambient temperature, V_{oc} is open circuit voltage, and SoC and SoH stand for state of charge and state of health, respectively.

1.6 New Control Paradigm

The fundamental idea behind the new control paradigm for next-generation WPT is to relocate all receiver-side controllers to the transmitter side while maintaining essential hardware components on the receiver side, as depicted in Figure 1.10. This approach eliminates the need for a communication channel and a DC-DC converter on the receiver side, thereby potentially

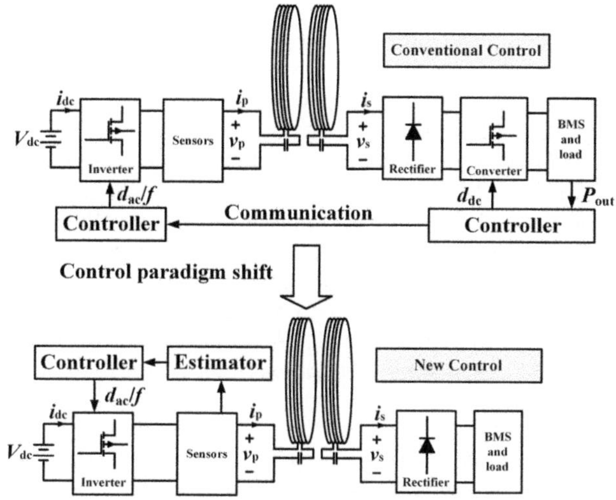

FIGURE 1.10
Paradigm shift in control for the next-generation WPT. (Adapted from reference [41], used with permission.)

improving reliability, cost-effectiveness, and power density of the receiver. By transferring control functions to the transmitter side, transmitter manufacturers assume responsibility for optimal control. Receiver manufacturers then focus solely on critical protective functions such as short-circuit and over-temperature protection using battery management systems (BMS) for battery loads. This shift promises substantial benefits for manufacturers of consumer electronics, medium-power devices, and EVs. The new control paradigm not only enhances WPT system performance but also preserves compatibility and interoperability.

It's important to highlight that the estimator (or digital twin) plays a crucial role in ensuring stability and achieving high performance in the new control paradigm. Front-end monitoring of receiver parameters and mutual couplings between the transmitter and receiver are fundamental tasks handled by the estimator. Throughout this book, various monitoring schemes will be discussed in detail in the subsequent chapters.

References

1. R. Lomas, *The Man Who Invented the Twentieth Century – Nikola Tesla – Forgotten Genius of Electricity*. London, UK: Headline Book Publishing, 1999.
2. N. Tesla, "Transmission of electrical energy without wire," *Electr. World Eng.*, vol. 1, pp. 21–24, Mar. 5, 1904. https://teslauniverse.com/nikola-tesla/articles/transmission-electric-energy-without-wires.

3. G. B. Joung and B. H. Cho, "An energy transmission system for an artificial heart using leakage inductance compensation of transcutaneous transformer," *IEEE Trans. Power Electron.*, vol.13, no. 6, pp. 1013–1022, Nov. 1998.
4. G. Covic and J. T. Boys, "Inductive power transfer," *Proc. IEEE [Invited Pap.]*, vol. 101, no. 6, pp. 1276–1289, Jun. 2013.
5. S. Y. R. Hui, "Planar wireless charging technology for portable electronic products and Qi," *Proc. IEEE [Invited Pap.]*, vol. 101, no. 6, pp. 1290–1301, Jun. 2013.
6. C. T. Rim and C. Mi, *Wireless Power Transfer*. New York: Wiley-IEEE Press, 2017.
7. Y. Jang and M. M. Jovanovic, "A contactless electrical energy transmission system for portable-telephone battery chargers," in *22nd International Telecommunications Energy Conference*, Phoenix, AZ, USA, 2000, pp. 726–772.
8. G. Roberts, A. Owens, P. Lane, et al., "A contactless transfer device for power and data," in *IEEE Aerospace Applications Conference Proceedings*, Aspen, CO, USA, 1996, pp. 333–345.
9. H. Brockmann and H. Turtiainen, "Charger with inductive power transmission for batteries in a mobile electrical device," U.S. Patent 6,118,249A, Sept. 12, 1999.
10. K. Oguri, "Power supply coupler for battery charger," U.S. Patent 6,356,049B2, Mar. 12, 2000.
11. C.-G. Kim, D.-H. Seo, J.-S. You, et al., "Design of a contactless battery charger for cellular phone," *IEEE Trans. Ind. Electron.*, vol. 48, no. 6, pp. 1238–1247, Dec. 2001.
12. T. Bieler, M. Perrottet, V. Nguyen, and Y. Perriard, "Contactless power and information transmission," *IEEE Trans. Ind. Appl.*, vol. 38, no. 5, pp. 1266–1272, Sep.–Oct. 2002.
13. S. Y. Hui, H. S. Chung, and S. C. Tang, "Coreless printed circuit board (PCB) transformers for power MOSFET/IGBT gate drive circuits," *IEEE Trans. Power Electron.*, vol. 14, no. 3, pp. 422–430, May 1999.
14. B. Choi, J. Nho, H. Cha, et al., "Design and implementation of low-profile contactless battery charger using planar printed circuit board windings as energy transfer device," *IEEE Trans. Ind. Electron.*, vol. 51, no. 1, pp. 140–147, Feb. 2004.
15. S. Y. R. Hui and W.C. Ho, "A new generation of universal contactless battery charging platform for portable Consumer Electronic equipment," *IEEE Trans. Power Electron.*, vol. 20, no. 3, pp. 620–627, May 2005.
16. L. Cheng, J. W. Hay, and P. Beart, "Contact-less power transfer," US Patent 6,906,495B2, Jun. 14, 2005.
17. Qi Specification [Online]: https://www.wirelesspowerconsortium.com/knowledge-base/specifications/download-the-qi-specifications/.
18. A. Kurs, A. Karalis, R. Moffatt, et al., "Wireless power transfer via strongly coupled magnetic resonance," *Science*, vol. 317, no. 5837, pp. 83–86, Jun. 2007.
19. AirFuel Resonant [Online]: https://airfuel.org/airfuel-resonant/.
20. SAE TIR J2954 [Online]: https://standards.sae.org/wip/j2954/.
21. Ki Standard [Online]: https://www.wirelesspowerconsortium.com/standards/ki-cordless-kitchen/.
22. LEV Standard [Online]: https://www.wirelesspowerconsortium.com/standards/light-electric-vehicles/.
23. Qi v2.0 [Online]: https://www.chargewithqi.com/media/w0ha5cbk/qi2-certification-rolls-out-news-release-11132023.pdf.

24. S. Y. R. Hui, W. X. Zhong, and C. K. Lee, "A critical review of recent progress in mid-range wireless power transfer," *IEEE Trans. Power Electron.*, vol. 29, no. 9, pp. 4500–4511, Sept. 2014.
25. H. Liang, Y. Yang, and S. Y. R. Hui, "Improvement of lithium-ion battery charging from the state-of-the-art industrial JEITA guidelines to a hybrid temperature-regulated current control," *IEEE Trans. Power Electron.*, vol. 37, no. 6, pp. 6412–6423, Jan. 2021.
26. M. Fu, H. Yin, X. Zhu, and C. Ma, "Analysis and tracking of optimal load in wireless power transfer systems," *IEEE Trans. Power Electron.*, vol.30, no.7, pp. 3952–3963, Jul. 2015.
27. H. Li, J. Li, K. Wang, et al., "A maximum efficiency point tracking control scheme for wireless power transfer systems using magnetic resonant coupling," *IEEE Trans. Power Electron.*, vol. 30, no. 7, pp. 3998–4008, Jul. 2015.
28. Y. Yang, W. Zhong, S. Kiratigongvoot, et al., "Dynamic improvement of series-series compensated wireless power transfer systems using discrete sliding mode control," *IEEE Trans. Power Electron.*, vol.33, no.7, pp. 6351–6360, Jul. 2018.
29. W. X. Zhong and S. Y. R. Hui, "Maximum energy efficiency tracking for wireless power transfer systems," *IEEE Trans. Power Electron.*, vol. 30, no. 7, pp. 4025–4034, Jul. 2015.
30. W. Zhong and S. Y. R. Hui, "Charging time control of wireless power transfer systems without using mutual coupling information and wireless communication system," *IEEE Trans. Ind. Electron.*, vol. 64, no. 1, pp. 228–235, Jan. 2017.
31. H. Ishihara, F. Moritsuka, H. Kudo, et al., "A voltage ratio-based efficiency control method for 3kW wireless power transmission," in *IEEE Applied Power Electronics Conference and Exposition*, Fort Worth, Tx, USA, Mar. 2014, pp. 1312–1316.
32. F. Pellitteri, V. Boscaino, R. Miceli, and U. K. Madawala, "Power tracking with maximum efficiency for wireless charging of E-bikes," in *2014 IEEE International Electric Vehicle Conference (IEVC)*, Florence, Italy, Dec. 2014, pp. 1–1.
33. M. Budhia, G. A. Covic, and J. T. Boys, "Design and optimization of circular magnetic structures for lumped inductive power transfer systems," *IEEE Trans. Power Electron.*, vol. 26, no. 11, pp. 3096–3108, Sep. 2011.
34. M. Fu, C. Ma, and X. Zhu, "A cascaded boost-buck converter for high efficiency wireless power transfer systems," *IEEE Trans. Ind. Inform.*, vol. 10, no. 3, pp. 1972–1980, Aug. 2014.
35. R. Bosshard, J. W. Kolar, and B. Wunsch, "Control method for inductive power transfer with high partial-load efficiency and resonance tracking," in *2014 International Power Electronics Conference (ECCE-Asia)*, Hiroshima, Japan, 2014, pp. 2167–2174.
36. F. Xu, S. C. Wong, and C. K. Tse, "Inductive power transfer system with maximum efficiency tracking control and real-time mutual inductance estimation," *IEEE Trans. Power Electron.*, vol.37, no.5, pp. 6156–6167, May 2022.
37. Z. Huang, S. C. Wong, and C. K. Tse, "An inductive-power-transfer converter with high efficiency throughout battery-charging process," *IEEE Trans. Power Electron.*, vol. 34, no. 10, pp. 10245–10255, Oct. 2019.
38. R. Mai, Y. Liu, Y. Li, et al., "An active-rectified-based maximum efficiency tracking method using an additional measurement coil for wireless power transfer," *IEEE Trans. Power Electron.*, vol. 33, no. 1, pp. 716–728, Jan. 2018.

39. H.-W. R. Liang, Y. Yang, L. He, et al., "A multi-hysteresis control for minimizing battery charging time within industrial JEITA guidelines," *IEEE Trans. Ind. Electron.*, vol. 70, no. 8, pp. 8416–8425, Aug. 2023.
40. Y. Jiang, J. Yang, Z. Sun, K. Wang, and Y. Yang, "A secured fast charging control based on a machine learning-aided electrothermal model for lithium-ion batteries," in *IEEE Energy Conversion and Exposition (ECCE)*, Phoenix, Arizona, USA, 2024, pp. 1–8.
41. S. Y. R. Hui, Y. Yang, and C. Zhang, "Wireless power transfer: A paradigm shift for the next generation," *IEEE J. Emerg. Sel. Top. Power Electron*, vol. 11, no. 3, pp. 2412–2427, Jun. 2023.

2

Overview of the New Control Paradigm

2.1 Essential Challenge of the New Control Paradigm

The new control paradigm is characterized by three key elements: (i) transmitter-side control, (ii) communication-free interaction between transmitters and receivers, and (iii) optimal performance. While each element individually is not new, the integration of the three components presents challenges that have only recently been addressed through the technologies we have developed. Before we present our technologies in this book, let's take a moment to acknowledge the foundational work of the pioneers whose innovations we build upon.

An early implementation of transmitter-side control is detailed in [1], presenting a direct feedback control for a typical WPT system with an inductance-capacitance-inductance (LCL)-compensated transmitter resonator and an LC-compensated receiver resonator. The equivalent circuit and the corresponding control block diagram are plotted, as shown in Figure 2.1.

Based on the output voltage of the system V_o and the current I_1 through inductor L_1 in the transmitter-side LCL circuit, the voltage v_{c1} across capacitor C_1 can be expressed using the following equation:

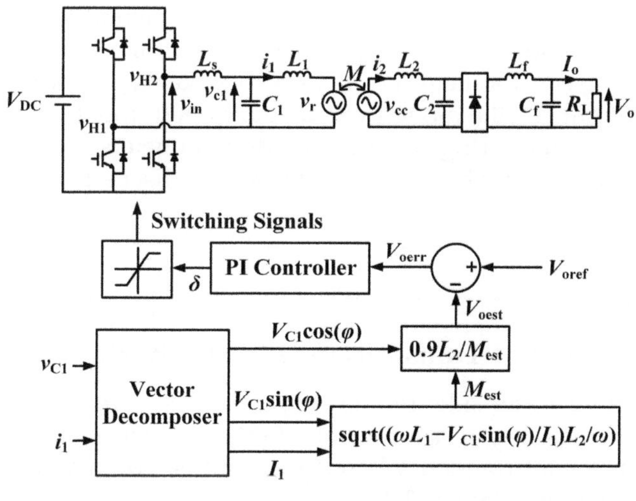

FIGURE 2.1
Control block diagram of the classic direct feedback control for a typical two-stage WPT system.

DOI: 10.1201/9781003468523-2

$$v_{c1} = \frac{\pi M}{2\sqrt{2}L_2} V_o + j\left(\omega L_1 - \frac{\omega M^2}{L_2}\right) I_1 \tag{2.1}$$

where M is the mutual inductance between the coupled coils, L_1 and L_2 are the self-inductances of the transmitter-side and receiver-side coils, and ω is the switching angular frequency. By decoupling the real and imaginary parts in Equation (2.1), the following equations can be derived:

$$M = \sqrt{\left(L_1 - \frac{V_{c1}\sin(\varphi)}{\omega I_1}\right) L_2} \tag{2.2}$$

$$V_o = \frac{2\sqrt{2}V_{c1}\cos(\varphi)L_2}{\pi M} \tag{2.3}$$

where φ represents the phase difference between v_{c1} and i_1. Equation (2.2) allows for the estimation of mutual inductance based on the measurements from the transmitter side. This estimated mutual inductance can then be utilized to calculate the output voltage V_o. However, the estimated mutual inductance using this method is highly sensitive to the measured system parameters when operating at the resonant frequency. As a result, the accuracy of the output voltage control cannot be ensured if the sensor precision is low. Similar transmitter-side control issues are present in [2–5]. The primary challenge arises from accurately estimating mutual inductance or other receiver-side parameters in the absence of communication feedback. It follows that developing perfect digital twins of receivers could address this accuracy issue. For the new control paradigm aimed at achieving optimal performance, this remains a critical concern that must be resolved. By stripping away unnecessary details, we can conclude that the essential technology needed for the success of the new control paradigm is the establishment of digital twins for the receivers in WPT systems. To address hardware and cost constraints, elite models can be developed initially. Consequently, transmitter-side monitoring of receiver-side parameters, including mutual coupling conditions, will be an important topic of discussion in this book.

2.2 Development of Transmitter-Side Monitoring

Early-stage transmitter-side monitoring strategies are developed using non-resonant frequency measurements [6–16]. The schematic diagram of a typical non-resonant frequency-based transmitter-side monitoring scheme is plotted, as shown in Figure 2.2.

The basic idea behind these non-resonant frequency strategies is to derive the necessary circuit equations for the system at various operating

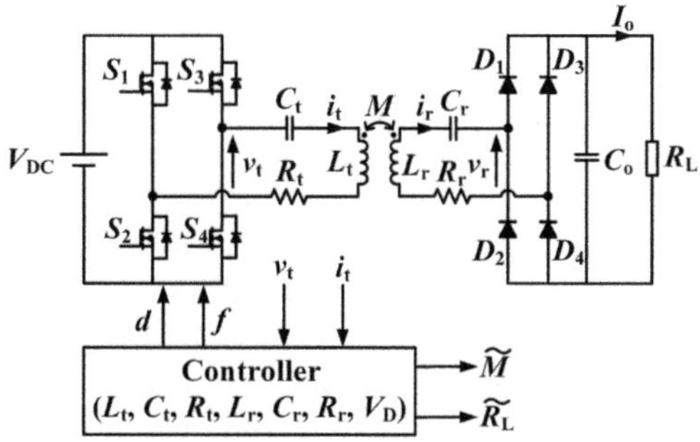

FIGURE 2.2
Schematic diagram of a typical non-resonant frequency-based transmitter-side monitoring strategy.

frequencies, in order to identify key variable parameters, such as the mutual inductances between the coupled coils and the load resistances. Theoretically, only two circuit equations are required to determine the mutual inductance and the load resistance. Given the decoupling of the real and imaginary parts in the equation, it is theoretically sufficient to measure the system parameters at a single frequency. However, the presence of parasitic parameters and measurement errors introduces significant inaccuracies. To reduce these random errors and improve accuracy, it is generally necessary to take measurements across multiple frequencies to ensure reliable monitoring. The advantages of employing non-resonant frequency measurements include the availability of multiple datasets across a broad frequency spectrum and lower measurement sensitivities compared to measurements taken at resonant frequencies. In [6], the load conditions of domino SS-compensated WPT systems are effectively monitored under the assumption of purely resistive loads. Similar assumptions are made in [7,8] to monitor the mutual inductance of a two-stage SS-compensated WPT system and the load conditions of WPT systems with multiple outputs. Sensitivity analyses demonstrate that the estimation equations derived from non-resonant frequency measurements can provide greater accuracy than those based on resonant frequency measurements. An improved monitoring scheme utilizing heuristic algorithms is presented in [9], yielding more accurate results even with unknown receiver-side parameters. Additionally, [10] introduces an enhanced monitoring strategy based on hardware load shunting, addressing the issue of non-resistive load characteristics commonly encountered in practice, which can diminish monitoring accuracy. Similar advanced monitoring approaches that account for unknown system parameters or non-resistive load characteristics are discussed in [11–16]. However, all non-resonant frequency methods conflict with the maximum efficiency operation of WPT systems, which require maintaining constant output currents, a critical factor for long-duration battery charging. These methods are also susceptible to parameter drifts or disturbances that are not accounted for in the estimation equations.

Most estimation equations based on non-resonant frequency methods are highly dependent on system parameters, which means the system parameters must be accurately known in advance. In practice, some parameters, like the self-inductances of coils with ferrites, may vary. Therefore, those conventional non-resonant frequency-based monitoring strategies are only suitable for specific applications. General front-end monitoring strategies do not truly exist. This is one of the reasons I wrote this book: to provide guidance on selecting the most appropriate method for future practical applications. I hope my efforts could be of value to you.

To facilitate front-end monitoring of mutual coupling and load conditions in WPT systems operating at resonant frequencies, several methods based on three distinct approaches have been proposed. The first approach is to use dynamic response information to estimate load conditions [17,18]. In [17], the transmitter-side inverter delivers power to the transmitter-side resonator to maintain the amplitude of the transmitter-side current during the initial energy injection phase. Once the energy injection concludes, the system transitions to the free-resonant mode, where the primary resonant current begins to decay due to the lack of external power supplies. The load conditions are assessed by monitoring the decay rates of the transmitter-side currents. However, this method faces a significant challenge regarding the consistency of the power supply. For battery charging applications, fluctuations in power supply are unacceptable due to their impact on charging speed and the potential for battery aging. In [18], a dynamic front-end monitoring strategy based on resonant frequency and the small-signal model of the WPT system is proposed. This method allows for the estimation of mutual inductance and load conditions using only the dynamic response of the transmitter-side current, eliminating the need for communication feedback from the receiver. The schematic diagram of a typical dynamic response-based front-end monitoring strategy is plotted, as shown in Figure 2.3.

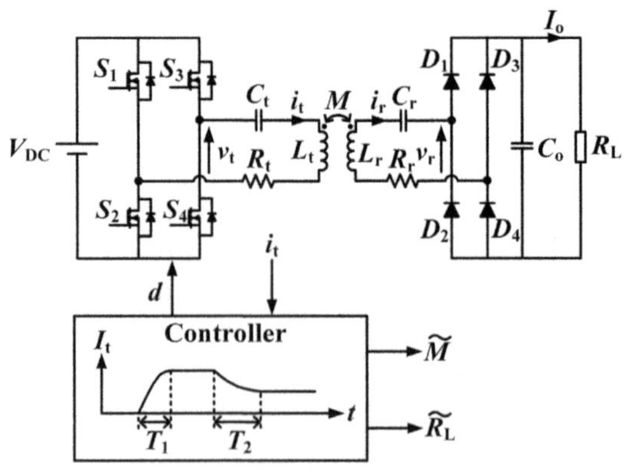

FIGURE 2.3
Schematic diagram of a typical dynamic response-based front-end monitoring strategy.

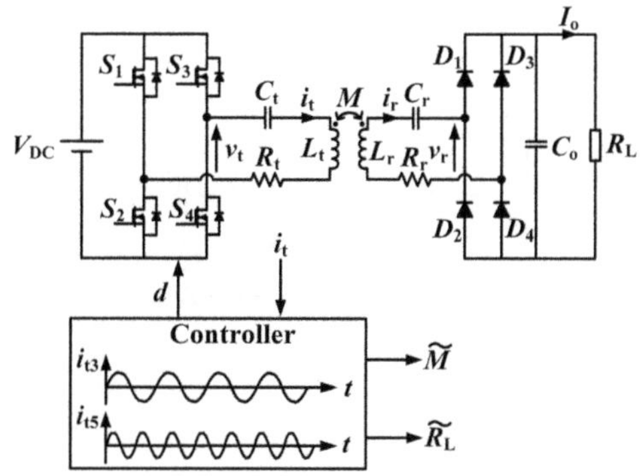

FIGURE 2.4
Schematic diagram of a typical harmonic-based front-end monitoring strategy.

A key advantage of this approach is its ability to accurately and swiftly determine mutual inductance and load conditions at the resonant frequency during the initial soft-start phase. However, this approach has limitations in accurately determining the settling time of the dynamics. The second approach leverages transmitter-side harmonic information to estimate load and coupling conditions [19]. While effective at resonant frequencies, this method faces challenges in accurately measuring higher-order harmonic components when the transmitter-side signals are weak due to certain coupling and load conditions. The schematic diagram of a typical harmonic-based front-end monitoring strategy is plotted, as shown in Figure 2.4.

The third approach uses AI algorithms to address uncertainties arising from parasitic parameters, measurement errors, and temperature fluctuations [20–22]. However, these AI-based methods require large datasets to achieve optimal performance. The schematic diagram of a typical AI-based front-end monitoring strategy is plotted, as shown in Figure 2.5.

2.3 From Transmitter-Side Monitoring to Transmitter-Side Control

Once the critical parameters, such as mutual inductance and load resistance, are successfully monitored, the system outputs can be controlled at the front end, without requiring communication feedback from the receiver to the transmitter. For instance, if the mutual inductance and load resistance of the SS-compensated WPT system in Figure 2.5 are monitored, the output current can be theoretically estimated at the transmitter side by

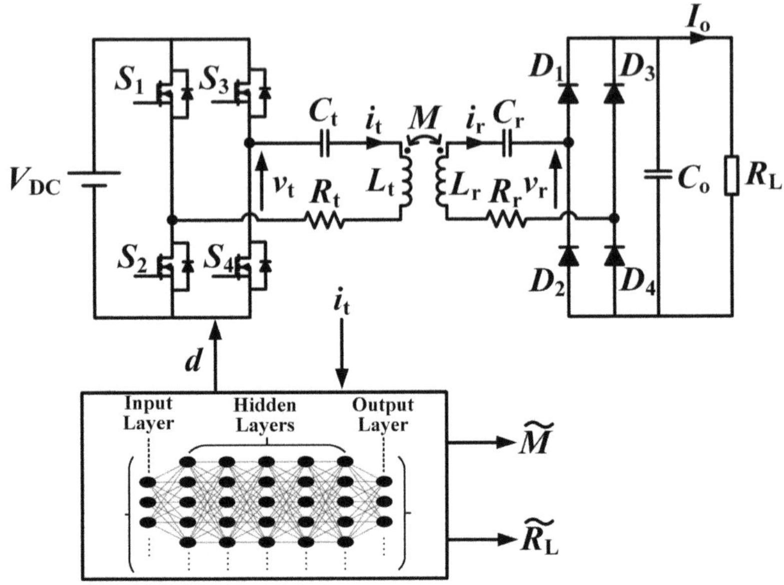

FIGURE 2.5
Schematic diagram of a typical AI-based front-end monitoring strategy.

$$I_o = \frac{\frac{8}{\pi^2}\omega_o \tilde{M} V_{DC}}{R_t R_r + \frac{8}{\pi^2} R_t \tilde{R}_L + \omega_o{}^2 \tilde{M}^2} \tag{2.4}$$

where $\omega_o = \frac{1}{\sqrt{L_t C_t}} = \frac{1}{\sqrt{L_r C_r}}$ is the resonant angular frequency. From Equation (2.4), we can identify two critical issues associated with transmitter-side monitoring-based output control that must not be overlooked. One issue is that monitoring errors in mutual inductance and load resistance can accumulate, leading to inaccuracies in the output control. The second issue is that accurate system parameters must be known in advance. However, this is challenging for many WPT systems operating over long periods, as temperature variations can affect the ESRs and parasitic components, such as diode forward voltages, which are not accounted for in Equation (2.4). Additionally, in WPT systems with ferrites, the self-inductances of the coupled coils vary with changes in current. This variation further deteriorates output accuracy. Therefore, a thorough sensitivity analysis of the estimated and variable parameters with respect to system outputs is necessary to ensure that the estimated equations can be applied in practice with high accuracies.

I must acknowledge that the accuracy of transmitter-side monitoring-based control strategies may be limited for certain WPT systems. To improve accuracy, I propose two potential strategies. The first is to enhance the monitoring accuracy of estimated parameters, such as mutual inductance

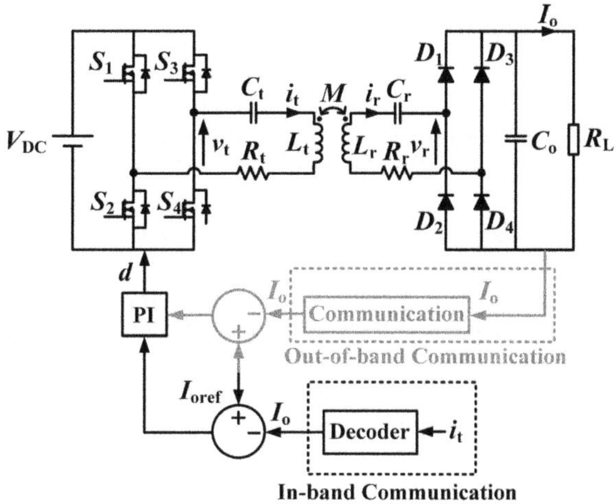

FIGURE 2.6
Schematic diagram of a typical WPT system with an in-band communication replacing the out-of-band communication.

and load resistance. The second is to compensate for output offsets caused by variable parameters, such as ESRs, diode forward voltages, and the self-inductances of the coupled coils. If accuracy cannot be sufficiently ensured, I recommend retaining the conventional feedback control structure while replacing the traditional (out-of-band) communication with in-band communication, as shown in Figure 2.6.

Amplitude shift keying (ASK) and frequency shift keying (FSK) in the "Qi" standard are common in-band communication techniques [23]. Similarly, the transmitter-side control in [5] utilizes an in-band communication approach. The "talkative" technique can also be regarded as an in-band communication method. In-band communication techniques, while capable of transmitting messages in real time from the receiver to the transmitter, face challenges such as high computational overhead for encoding and decoding, as well as susceptibility to in-band interference.

References

1. D. J. Thrimawithana and U. K. Madawala, "A primary side controller for inductive power transfer systems," in *2010 IEEE International Conference on Integrated Circuit Design and Technology*, Via del Mar, Chile, 2010, pp. 661–666.
2. C. Wang, C. Zhu, K. Song, et al., "Primary-side control method in two-transmitter inductive wireless power transfer systems for dynamic wireless charging applications," in *2017 IEEE PELS Workshop on Emerging Technologies: Wireless Power Transfer (WoW)*, Chongqing, China, 2017, pp. 1–6.

3. X. Meng, D. Qiu, B. Zhang, and W. Xiao, "Output voltage stabilization control without secondary side measurement for implantable wireless power transfer system," in *2018 IEEE PELS Workshop on Emerging Technologies: Wireless Power Transfer (Wow)*, Montreal, QC, Canada, 2018, pp. 1–5.
4. A. Alkasir, S. Ehsan Abdollahi, S. Reza Abdollahi, and P. Wheeler, "A primary side CCS-MPC controller for constant current/voltage charging operation of series-series compensated wireless power transfer systems," in *2021 12th Power Electronics, Drive Systems, and Technologies Conference (PEDSTC)*, Tabriz, Iran, 2021, pp. 1–5.
5. Y. Yang, J. Zeng, and S. Y. R. Hui, "A fast primary-side current and voltage control for direct wireless battery chargers," *IEEE J. Emerg. Sel. Topics Power Electron.*, vol. 12, no. 1, pp. 355–366, Feb. 2024.
6. J. Yin, D. Lin, C. K. Lee, and S. Y. R. Hui, "A systematic approach for load monitoring and power control in wireless power transfer systems without any direct output measurement," *IEEE Trans. Power Electron.*, vol. 30, no. 3, pp. 1657–1667, Mar. 2015.
7. J. Yin, D. Lin, T. Parisini, and S. Y. Hui, "Front-end monitoring of the mutual inductance and load resistance in a series–series compensated wireless power transfer system," *IEEE Trans. Power Electron.*, vol. 31, no. 10, pp. 7339–7352, Oct. 2016.
8. J. Yin, D. Lin, C. K. Lee, T. Parisini, and S. Y. Hui, "Front-end monitoring of multiple loads in wireless power transfer systems without wireless communication systems," *IEEE Trans. Power Electron.*, vol. 31, no. 3, pp. 2510–2517, Mar. 2016.
9. Y. Yang, S. -C. Tan, and S. Y. R. Hui, "Front-end parameter monitoring method based on two-layer adaptive differential evolution for SS-compensated wireless power transfer systems," *IEEE Trans. Ind. Informat.*, vol. 15, no. 11, pp. 6101–6113, Nov. 2019.
10. Y. Yang, S. C. Tan, and S. Y. R. Hui, "Fast hardware approach to determining mutual coupling of series–series-compensated wireless power transfer systems with active rectifiers," *IEEE Trans. Power Electron.*, vol. 35, no. 10, pp. 11026–11038, Oct. 2020.
11. J. P. W. Chow and H. S. H. Chung, "Use of primary-side information to perform online estimation of the secondary-side information and mutual inductance in wireless inductive link," in *IEEE Applied Power Electronics Conference and Exposition (APEC)*, Charlotte, NC, USA, 2015, pp. 2648–2655.
12. Y. Mao, Y. Yang, and K. Wang, "A primary-side fixed-frequency CC and CV output control for single-stage wireless battery chargers with series compensation receivers," *IEEE Trans. Transp. Electr.*, vol. 10, no. 2, pp. 3407–3415, Jun. 2024.
13. J. Zeng, S. Chen, Y. Yang, and S. Y. R. Hui, "A primary-side method for ultrafast determination of mutual coupling coefficient in milliseconds for wireless power transfer systems," *IEEE Trans. Power Electron.*, vol. 37, no. 12, pp. 15706–15716, Dec. 2022.
14. J. Zeng, J. Wu, K. Li, Y. Yang, and S. Y. R. Hui, "Dynamic monitoring of battery variables and mutual inductance for primary-side control of a wireless charging system," *IEEE Trans. Ind. Electron.*, vol. 71, no. 7, pp. 7966–7974, Jul. 2024.
15. J. Zeng, Y. Yang, K. Li, S. Chen, and S. Y. R. Hui, "An ultrafast estimation method for coupling coefficient and receiver resonant frequency in universal wireless power transfer systems," *IEEE Trans. Power Electron.*, vol. 39, no. 4, pp. 4870–4883, Apr. 2024.

16. K. Wang, J. Zeng, Y. Yang, and H. Xiao, "Advanced primary-side load monitoring schemes for inductive power transfer systems with enhanced nonlinear rectifier load models," in *IEEE Energy Conversion Congress and Exposition (ECCE)*, Nashville, TN, USA, 2023, pp. 1632–1639.
17. S. Lee, B. Choi, and C. T. Rim, "Dynamic characteristics of the inductive power transfer system for online electric vehicles by Laplace Phasor Transform," *IEEE Trans. Power Electron.*, vol. 28, no. 12, pp. 5902–5909, Dec. 2013.
18. Z. Sun, J. Zeng, and Y. Yang, "A resonant-frequency-based dynamic front-end monitoring strategy for series-series-compensated inductive power transfer systems without communication," *IEEE Trans. Ind. Appl.*, in press, 2025.
19. J. Zeng et al., "A fast front-end monitoring method for mutual inductance and load movement in SS-compensated wireless power transfer systems based on harmonics extraction," *IEEE Trans. Power Electron.*, vol. 40, no. 1, pp. 2569–2580, Jan. 2025.
20. Y. Yang, X. Zhang, and K. Wang, "A robust random forest regressor-based mutual inductance monitoring scheme for inductive power transfer systems," in *IEEE International Power Electronics and Application Conference and Exposition (PEAC)*, Guangzhou, China, Nov. 2022, pp. 171–176.
21. Y. Yang and H. Wu, "A robust primary-side hybrid data-driven load monitoring strategy for wireless power transfer systems" in *IEEE Energy Conversion Congress and Exposition (ECCE)*, Detroit, MI, USA, Oct. 2022, pp. 1–6.
22. K. Wang, Y. Yang, and X. Zhang, "Advanced front-end monitoring scheme for inductive power transfer systems based on random forest regression," in *IEEE Applied Power Electronics Conference and Exposition (APEC)*, Orlando, FL, USA, Mar. 2023, pp. 2901–2907.
23. Qi Specification [Online]: https://www.wirelesspowerconsortium.com/knowledge-base/specifications/download-the-qi-specifications/.

3

Non-Resonant Frequency-Based Transmitter-Side Monitoring Strategies

3.1 Fundamental Non-Resonant Frequency Method

3.1.1 Basic Concept

Non-resonant frequency methods are based on applying Kirchhoff's circuit laws to WPT systems operating at multiple non-resonant frequencies. The key advantage of using non-resonant frequencies is that transmitter-side monitoring of critical system parameters (such as mutual inductance and load conditions) is considerably less sensitive compared to systems operating at resonant frequencies. To illustrate the fundamental principles of the non-resonant frequency methods, we will use a typical SS-compensated WPT system as an example. The circuit diagram and its equivalent circuit in the fundamental frequency domain are depicted in Figure 3.1.

Here, v_t represents the input AC voltage of the coupled resonators, and ω denotes the switching angular frequency of the resonators. L_t and L_r are the self-inductances of the coupled coils, while R_t and R_r are the ESRs of the

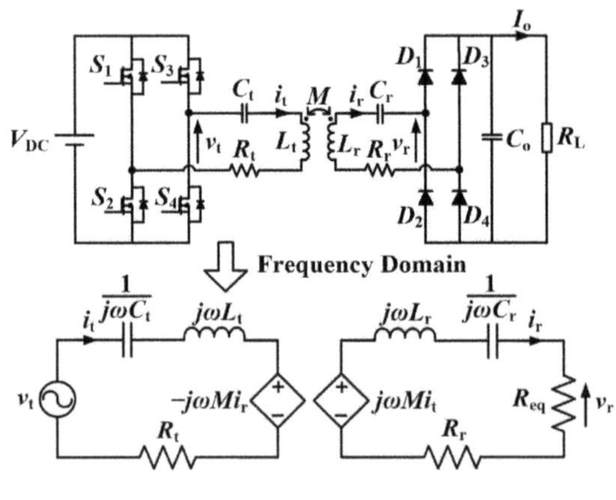

FIGURE 3.1
Schematic diagram and equivalent circuit in the fundamental frequency domain of a typical SS-compensated WPT system.

 DOI: 10.1201/9781003468523-3

resonators. C_t and C_r are the compensated capacitances, designed to satisfy the resonant condition $\omega_o = \frac{1}{\sqrt{L_t C_t}} = \frac{1}{\sqrt{L_r C_r}}$, where ω_o is the resonant angular frequency of the resonators. M is the mutual inductance between the coupled coils. R_{eq} is the equivalent resistance of the nonlinear rectifier load, which includes the diode-bridge rectifier, the filter capacitor C_o, and the output resistive load R_L. For most ideal nonlinear loads, $R_{eq} = \frac{8}{\pi^2} R_L$ [1]. Based on Kirchhoff's circuit laws,

$$v_t = \left(j\omega L_t + \frac{1}{j\omega C_t} + R_t \right) i_t - j\omega M i_r \tag{3.1}$$

$$j\omega M i_t = \left(j\omega L_r + \frac{1}{j\omega C_r} + R_r + R_{eq} \right) i_r \tag{3.2}$$

By combining Equations (3.1) and (3.2) and eliminating the term i_r, the relationship can be derived for cases where the switching angular frequency ω differs from the resonant angular frequency ω_o (i.e., $\omega \neq \omega_o$),

$$R_{eq} = \frac{\left(\omega L_r - \frac{1}{\omega C_r} \right) (Z_{in} \cos\theta - R_t)}{\omega L_t - \frac{1}{\omega C_t} - Z_{in} \sin\theta} - R_r \tag{3.3}$$

$$M = \sqrt{\frac{\left(\omega L_r - \frac{1}{\omega C_r} \right) \left[(Z_{in} \cos\theta - R_t)^2 + \left(\omega L_t - \frac{1}{\omega C_t} - Z_{in} \sin\theta \right)^2 \right]}{\omega^2 \left(\omega L_t - \frac{1}{\omega C_t} - Z_{in} \sin\theta \right)}} \tag{3.4}$$

where Z_{in} represents the input impedance of the coupled resonators, and θ is the phase difference between v_t and i_t (i.e., $Z_{in} \cos\theta = \mathbf{Re}\left(\frac{v_t}{i_t} \right)$ and $Z_{in} \sin\theta = \mathbf{Im}\left(\frac{v_t}{i_t} \right)$). The equivalent resistance R_{eq} and the mutual inductance M can be estimated using Equations (3.3) and (3.4), based solely on transmitter-side measurements (i.e., Z_{in} and θ). However, when the switching angular frequency equals the resonant angular frequency (i.e., $\omega = \omega_o$),

$$Z_{in} = R_t + \frac{\omega_o^2 M^2}{R_r + R_{eq}} \tag{3.5}$$

Since R_{eq} and M are interdependent in Equation (3.5), they cannot be estimated simultaneously when both parameters are unknown. Therefore, basic monitoring techniques on the transmitter side, particularly at non-resonant frequencies, are based on fundamental circuit principles. In theory, R_{eq} and M can be estimated in real time using the measured Z_{in} and θ at a specific non-resonant angular frequency ω, but this is only possible if the system parameters L_r, R_r, R_t, and ω_o are known beforehand. Subsequently, the load resistance R_L can be determined using the relation $R_L = \frac{\pi^2}{8} R_{eq}$. However, in practice, transmitter-side monitoring proves to be highly inaccurate due to several contributing factors:

1. The self-inductances of the transmitting and receiving coils can fluctuate as a result of magnetic saturation.
2. The ESRs of the transmitter and receiving coils are subject to variations due to changes in current strength and temperature. In fact, the AC resistance within the WPT system is highly sensitive to fluctuations induced by both current and temperature.
3. Practical components, such as compensating capacitances, are limited to discrete values and exhibit inherent tolerances, leading to inaccuracies in Equations (3.3) and (3.4).
4. Inaccuracies in measuring v_t and i_t can introduce errors in the derived parameters Z_{in} and θ, thereby compromising the precision of Equations (3.3) and (3.4).

Moreover, parasitic components in power diodes and rectifier cables, such as forward voltages and line resistances, can significantly affect the accuracy of the estimation equations and the relationship $R_L = \frac{\pi^2}{8} R_{eq}$. This is especially critical in low-power WPT systems operating at low output voltages. Furthermore, load conditions have a substantial impact on the precision of the relationship $R_L = \frac{\pi^2}{8} R_{eq}$, as highlighted in [1].

Figure 3.2 compares the measured and calculated values of R_L and M for a practical SS-compensated WPT system, with the nominal electrical parameters listed in Table 3.1. Significant discrepancies between the measured and calculated values are evident, largely attributable to the key factors discussed earlier. To identify the dominant contributing factors, one effective approach is to isolate each factor by keeping others constant during experimentation and systematically varying one factor at a time. If experimental validation is not feasible, a sensitivity analysis of the critical parameters can be conducted to assess their individual contributions and impacts.

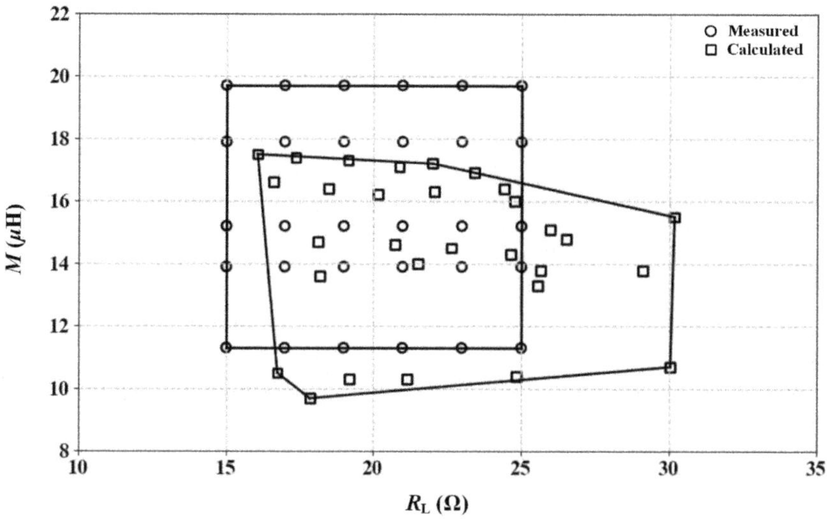

FIGURE 3.2
Comparisons between the measured and calculated load resistance and mutual inductance of a practical SS-compensated WPT system.

TABLE 3.1

Nominal Electrical Parameters of a Practical SS-Compensated WPT System

Parameters	Value	Parameters	Value
Transmitter inductance L_t	49.01 μH	Receiver inductance L_r	48.66 μH
Transmitter capacitance C_t	51.91 nF	Receiver capacitance C_r	51.81 nF
Transmitter resistance R_t	0.283 Ω	Receiver resistance R_r	0.225 Ω
Resonant frequency f_o	99.8 kHz	Operating frequency f	110 kHz
Input DC voltage V_{DC}	10 V	Output DC capacitance C_o	812 nF
Diode forward voltage V_D	1.5 V		

3.1.2 Sensitivity Analysis

Examples of sensitivity analysis applied to typical WPT systems can be found in [2,3]. In this book, I will perform a sensitivity analysis on the critical parameters influencing the estimation of load resistance and mutual inductance, focusing on the SS-compensated WPT system depicted in Figure 3.1. The methodology for this analysis involves calculating the partial derivatives of the load resistance or mutual inductance with respect to the critical parameters. This approach aligns with the core principles of sensitivity analysis, enabling an evaluation of how variations in the critical parameters impact the monitored parameters.

For the SS-compensated WPT system illustrated in Figure 3.1, the partial derivatives of the estimated load resistance and mutual inductance with respect to the self-inductances of transmitter and receiving coils can be derived from Equation (3.3), the relationship $R_L = \frac{\pi^2}{8} R_{eq}$, and Equation (3.4) as follows:

$$\frac{\partial R_L}{\partial L_t} = \frac{\omega^3 M^2 (Z_{in}\cos\theta - R_t)}{\left(Z_{in}\sin\theta - \omega L_t + \frac{1}{\omega C_t}\right)\left[(Z_{in}\cos\theta - R_t)^2 + \left(Z_{in}\sin\theta - \omega L_t + \frac{1}{\omega C_t}\right)^2\right]} \quad (3.6)$$

$$\frac{\partial M}{\partial L_t} = -\frac{\omega M\left[\left(Z_{in}\sin\theta - \omega L_t + \frac{1}{\omega C_t}\right)^2 - (Z_{in}\cos\theta - R_t)^2\right]}{2\left(Z_{in}\sin\theta - \omega L_t + \frac{1}{\omega C_t}\right)\left[(Z_{in}\cos\theta - R_t)^2 + \left(Z_{in}\sin\theta - \omega L_t + \frac{1}{\omega C_t}\right)^2\right]} \quad (3.7)$$

$$\frac{\partial R_L}{\partial L_r} = -\frac{\omega(Z_{in}\cos\theta - R_t)}{Z_{in}\sin\theta - \omega L_t + \frac{1}{\omega C_t}} \quad (3.8)$$

$$\frac{\partial M}{\partial L_r} = -\frac{(Z_{in}\cos\theta - R_t)^2 + \left(Z_{in}\sin\theta - \omega L_t + \frac{1}{\omega C_t}\right)^2}{2\omega M\left(Z_{in}\sin\theta - \omega L_t + \frac{1}{\omega C_t}\right)} \quad (3.9)$$

The sensitivity analysis curves of L_t and L_r with respect to the load resistance and mutual inductance estimations of the system, using the parameters provided in Table 3.1, are plotted based on Equations (3.6), (3.7), (3.8), and (3.9), as shown in Figures 3.3 and 3.4, respectively. Both figures clearly show that the sensitivities of load resistance and mutual inductance to the self-inductances of the transmitter and receiving coils are significantly high near the resonant frequencies. This indicates that small deviations of the self-inductances of the transmitter and receiving coils would result in large estimation errors. Therefore, it is recommended to estimate the load resistance and mutual inductance using measurements taken at frequencies well away from the resonant frequency.

A similar sensitivity analysis of the ESRs on the load resistance and mutual inductance estimations can be conducted using the partial derivative equations derived from Equations (3.3) and (3.4), as outlined below:

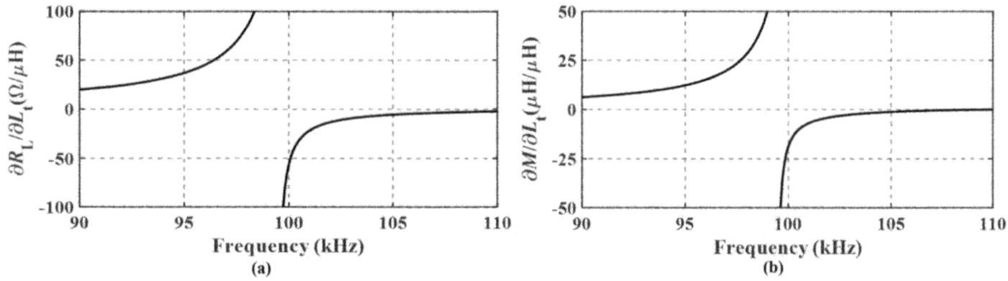

FIGURE 3.3
Sensitivity analysis curve of the self-inductance of the transmitter coil with respect to the (a) load resistance and (b) mutual inductance estimations of the WPT system.

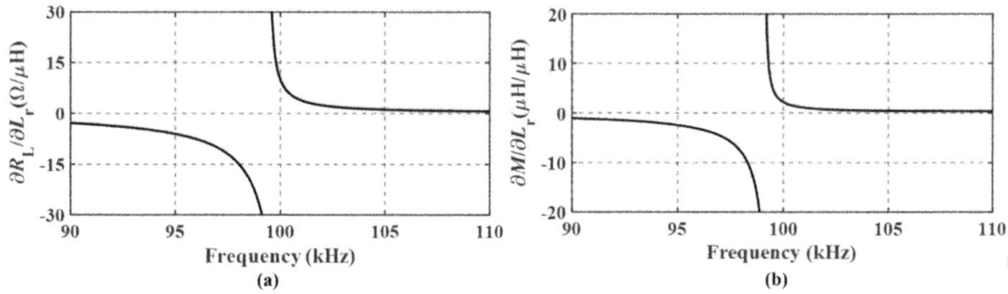

FIGURE 3.4
Sensitivity analysis curve of the self-inductance of the receiving coil with respect to the (a) load resistance and (b) mutual inductance estimations of the WPT system.

$$\frac{\partial R_L}{\partial R_t} = -\frac{\omega^2 M^2}{\left(Z_{in}\cos\theta - R_t\right)^2 + \left(Z_{in}\sin\theta - \omega L_t + \dfrac{1}{\omega C_t}\right)^2} \tag{3.10}$$

$$\frac{\partial M}{\partial R_t} = -\frac{M\left(Z_{in}\cos\theta - R_t\right)}{\left(Z_{in}\cos\theta - R_t\right)^2 + \left(Z_{in}\sin\theta - \omega L_t + \dfrac{1}{\omega C_t}\right)^2} \tag{3.11}$$

$$\frac{\partial R_L}{\partial R_r} = -1 \tag{3.12}$$

$$\frac{\partial M}{\partial R_r} = 0 \tag{3.13}$$

The sensitivity analysis curves of R_t and R_r with respect to R_L and M are shown in Figures 3.5 and 3.6, respectively. For the WPT system with the parameters provided in Table 3.1, deviations in R_t have a diminishing impact

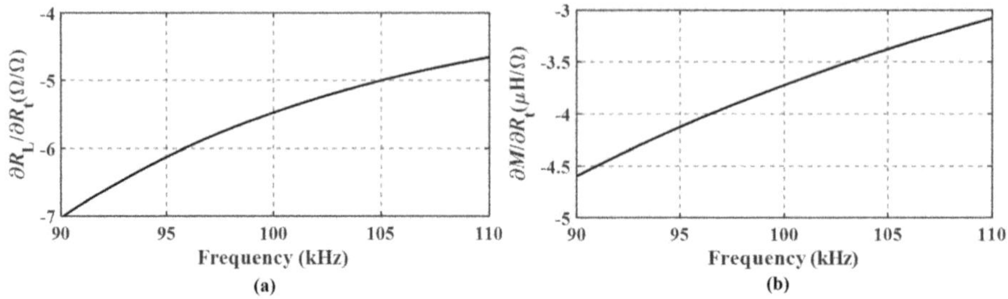

FIGURE 3.5
Sensitivity analysis curve of the ESR of the transmitter coil with respect to the (a) load resistance and (b) mutual inductance estimations of the WPT system.

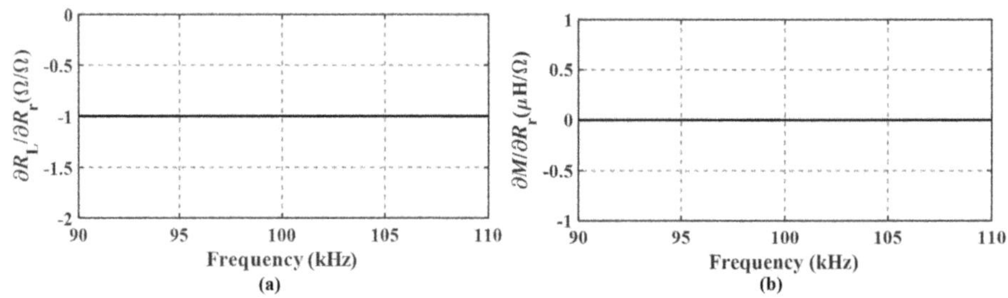

FIGURE 3.6
Sensitivity analysis curve of the ESR of the receiving coil with respect to the (a) load resistance and (b) mutual inductance estimations of the WPT system.

on the estimation accuracy of mutual inductance and load resistance as the operating frequencies increase. Therefore, it is recommended to determine the mutual inductance and load resistance of this WPT system at non-resonant frequencies above the resonant frequency. The partial derivatives of R_L and M with respect to R_r are constants, indicating that deviations in R_r do not influence the estimation accuracies of R_L and M.

Deviations in the practical compensated capacitances also impact the estimation accuracy of load resistance and mutual inductance, as demonstrated by the sensitivity analysis curves in Figures 3.7 and 3.8. The sensitivities are relatively low at frequencies significantly removed from the resonant frequency. These curves are derived based on the following partial derivative equations:

$$\frac{\partial R_{\mathrm{L}}}{\partial C_{\mathrm{t}}}=\frac{\omega M^2(Z_{\mathrm{in}}\cos\theta-R_{\mathrm{t}})}{C_{\mathrm{t}}^2\left(Z_{\mathrm{in}}\sin\theta-\omega L_{\mathrm{t}}+\dfrac{1}{\omega C_{\mathrm{t}}}\right)\left[(Z_{\mathrm{in}}\cos\theta-R_{\mathrm{t}})^2+\left(Z_{\mathrm{in}}\sin\theta-\omega L_{\mathrm{t}}+\dfrac{1}{\omega C_{\mathrm{t}}}\right)^2\right]} \tag{3.14}$$

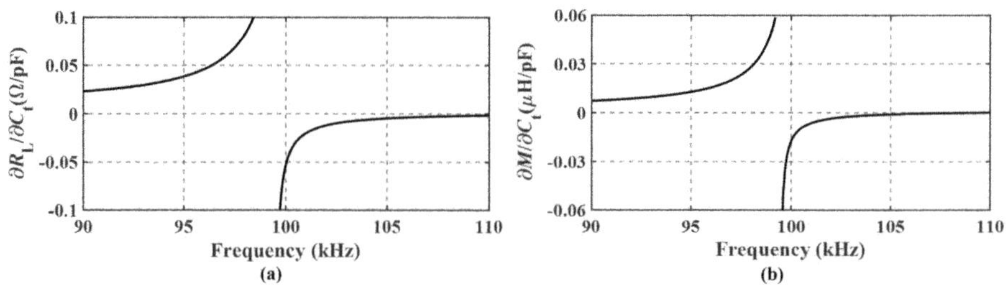

FIGURE 3.7
Sensitivity analysis curve of the transmitter compensated capacitance with respect to the (a) load resistance and (b) mutual inductance estimations of the WPT system.

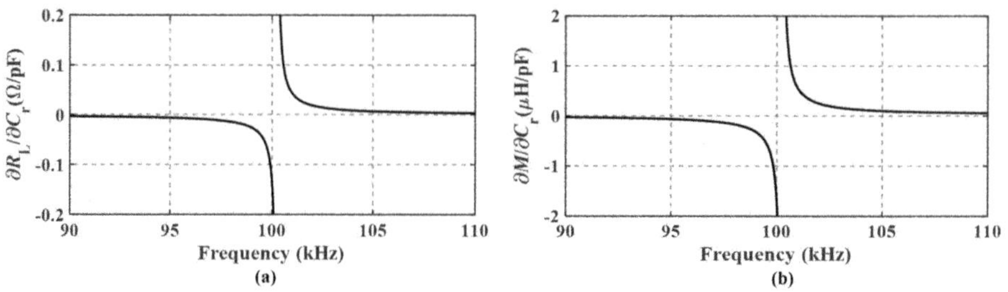

FIGURE 3.8
Sensitivity analysis curve of the receiving compensated capacitance with respect to the (a) load resistance and (b) mutual inductance estimations of the WPT system.

$$\frac{\partial M}{\partial C_t} = -\frac{M\left[-(Z_{in}\cos\theta - R_t)^2 + \left(Z_{in}\sin\theta - \omega L_t + \frac{1}{\omega C_t}\right)^2\right]}{2\omega C_t^2\left(Z_{in}\sin\theta - \omega L_t + \frac{1}{\omega C_t}\right)\left[(Z_{in}\cos\theta - R_t)^2 + \left(Z_{in}\sin\theta - \omega L_t + \frac{1}{\omega C_t}\right)^2\right]} \tag{3.15}$$

$$\frac{\partial R_L}{\partial C_r} = -\frac{Z_{in}\cos\theta - R_t}{\omega C_r^2\left(Z_{in}\sin\theta - \omega L_t + \frac{1}{\omega C_t}\right)} \tag{3.16}$$

$$\frac{\partial M}{\partial C_r} = -\frac{(Z_{in}\cos\theta - R_t)^2 + \left(Z_{in}\sin\theta - \omega L_t + \frac{1}{\omega C_t}\right)^2}{2M\omega^3 C_r^2\left(Z_{in}\sin\theta - \omega L_t + \frac{1}{\omega C_t}\right)} \tag{3.17}$$

The sensitivity analysis curves of Z_{in} and θ with respect to R_L and M are presented, derived from the following partial derivative equations, to illustrate the impact of transmitter-side measurement errors on the estimation errors of load resistance and mutual inductance:

$$\frac{\partial R_L}{\partial Z_{in}} = -\frac{\omega^2 M^2 \left[\left(\omega L_t - \frac{1}{\omega C_t}\right)\cos\theta - R_t \sin\theta\right]}{Z_1\left[(Z_{in}\cos\theta - R_t)^2 + Z_1^2\right]} \tag{3.18}$$

$$\frac{\partial M}{\partial Z_{in}} = -\frac{M\left[-2Z_1\,(Z_{in}\cos\theta - R_t)\cos\theta + (Z_{in}\cos\theta - R_t)^2\sin\theta - Z_1^2\sin\theta\right]}{2Z_1\left[(Z_{in}\cos\theta - R_t)^2 + Z_1^2\right]} \tag{3.19}$$

$$\frac{\partial R_L}{\partial \theta} = -\frac{\omega^2 M^2 Z_{in}\left[Z_{in} - \left(\omega L_t - \frac{1}{\omega C_t}\right)\sin\theta - R_t\cos\theta\right]}{Z_1\left[(Z_{in}\cos\theta - R_t)^2 + Z_1^2\right]} \tag{3.20}$$

$$\frac{\partial M}{\partial \theta} = -\frac{M Z_{in}\left[2(Z_{in}\cos\theta - R_t)Z_1\sin\theta + (Z_{in}\cos\theta - R_t)^2\cos\theta - Z_1^2\cos\theta\right]}{2Z_1\left[(Z_{in}\cos\theta - R_t)^2 + Z_1^2\right]} \tag{3.21}$$

To simplify the expressions, $Z_1 = Z_{in}\sin\theta - \omega L_t + \frac{1}{\omega C_t}$ is used. The curves in Figure 3.9 reveals that the sensitivity of Z_{in} to R_L and M estimations is lower at frequencies above the resonant frequency compared to those below it. The frequencies with the minimum sensitivities are slightly higher than the resonant frequency. Meanwhile, the curves in Figure 3.10 indicate that the frequencies with the minimum sensitivity for the phase difference θ occur above the resonant frequency.

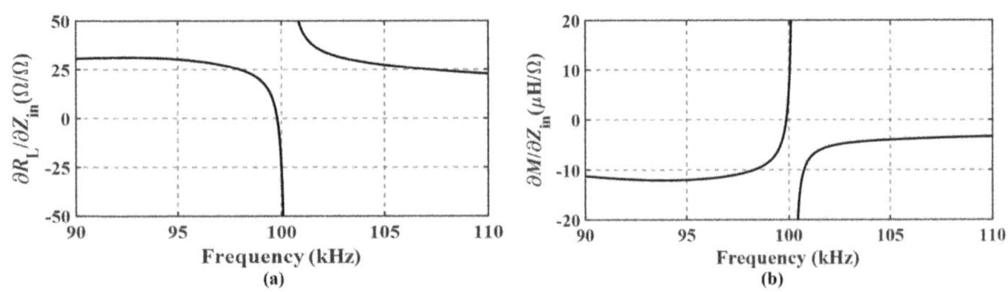

FIGURE 3.9
Sensitivity analysis curve of the input impedance Z_{in} with respect to the (a) load resistance and (b) mutual inductance estimations of the WPT system.

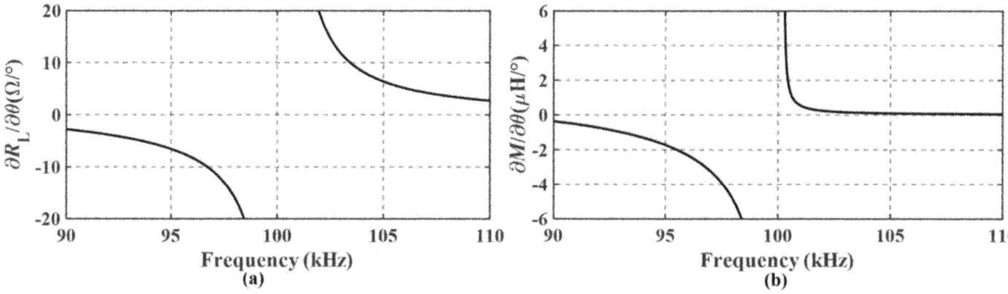

FIGURE 3.10
Sensitivity analysis curve of phase difference θ with respect to the (a) load resistance and (b) mutual inductance estimations of the WPT system.

As previously discussed, the non-resonant frequencies used to estimate load resistance and mutual inductance can vary significantly due to disturbances, uncertainties, or drifts in the system parameters. The simultaneous presence of system parameter uncertainties and measurement errors complicates the identification of the optimal non-resonant frequency in practice. For example, both the self-inductance and ESR of the transmitter coil tend to increase as the operating frequency changes. As a result, the accuracy of the fundamental non-resonant frequency method cannot be reliably guaranteed, given the deviations in various parameters during practical implementation.

3.2 Advanced Non-Resonant Frequency Methods

To streamline the process, advanced non-resonant frequency methods have been developed. In this book, I will present two approaches: a software-based non-resonant frequency method employing a heuristic algorithm and a hardware-based method utilizing an auxiliary circuit.

3.2.1 Heuristic Algorithm-Based Non-Resonant Frequency Method

The main objective of developing the heuristic algorithm-based non-resonant frequency method is to minimize the measurement errors inherent in the fundamental non-resonant frequency method, while simultaneously expanding the number of identifiable parameters. These parameters may encompass mutual inductance, load conditions of multiple receivers, and the characteristics of receiving resonators. The success of the heuristic algorithms in this context relies on their ability to perform global optimization rather than local optimization [4]. Commonly employed heuristic algorithms include Particle Swarm Optimization (PSO), Genetic Algorithm (GA), Differential Evolution (DE), and their variants.

Let me use an SS-compensated single-input multiple-output (SIMO) WPT system as an example to demonstrate the heuristic algorithm-based non-resonant frequency method. The circuit diagram and the equivalent circuit in the fundamental frequency domain of the SIMO-WPT system are illustrated in Figures 3.11 and 3.12, respectively.

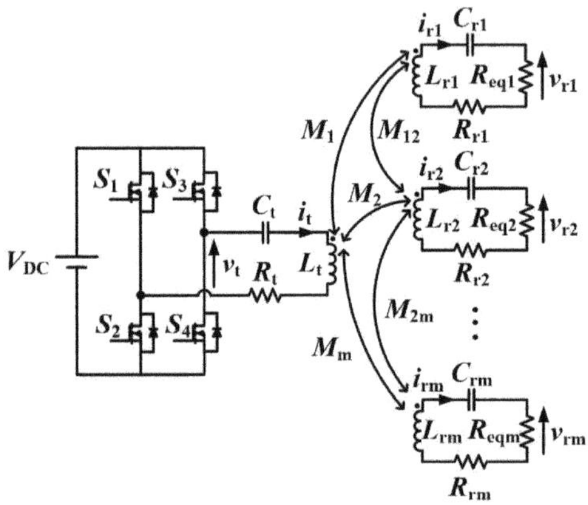

FIGURE 3.11
Schematic circuit diagram of a typical SS-compensated SIMO-WPT system.

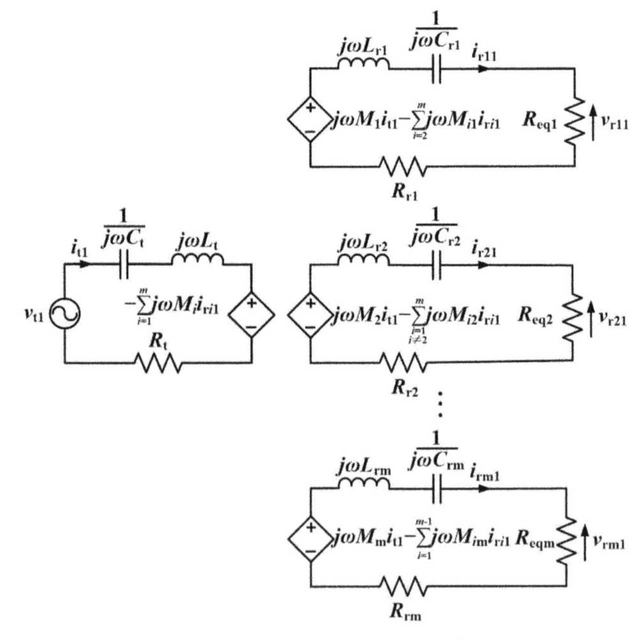

FIGURE 3.12
Equivalent circuit of the SS-compensated SIMO-WPT system in the fundamental frequency domain.

At the transmitter side, a full-bridge inverter is controlled to achieve a high-frequency AC voltage v_t. The transmitter resonator comprises a transmitter coil L_t, a compensated capacitor C_t, and their ESR R_t. On receiver sides, the receiving resonators comprise the coil inductances L_{ri} ($i = 1, 2, \ldots, m$), compensated capacitances C_{ri} ($i = 1, 2, \ldots, m$), their ESRs R_{ri} ($i = 1, 2, \ldots, m$), and equivalent load resistances R_{eqi} ($i = 1, 2, \ldots, m$). M_i ($i = 1, 2, \ldots, m$) are the mutual inductances between the transmitter and receiving coils. M_{ij} ($i = 1, 2, \ldots, m$, $j = 1, 2, \ldots, m$, $i \neq j$) are the mutual inductances between the receiving coils. The "self-mutual inductances" are physically null (i.e., $M_{11} = M_{22} = \cdots = M_{mm} = 0$). In the equivalent circuit, v_{t1}, i_{t1}, i_{r1}, and v_{r1} represent the fundamental components of v_t, i_t, i_r, and v_r, respectively. Based on the equivalent circuit, the following equations can be derived:

$$\begin{cases} v_{t1} = \left[R_t + \left(\omega L_t - \dfrac{1}{\omega C_t} \right) j \right] i_{t1} - \omega \mathbf{M}^T \mathbf{i}_{t1} j \\ \omega \mathbf{M} i_{t1} j - \omega \mathbf{M}_1 \mathbf{i}_{r1} j = \mathbf{Z}_r \mathbf{i}_{r1} \end{cases} \tag{3.22}$$

where $\mathbf{i}_{r1} = [i_{r11}, i_{r12} \ldots, i_{r1m}]^T$, $\mathbf{M} = [M_1, M_2 \ldots, M_m]^T$,

$$\mathbf{M}_1 = \begin{bmatrix} 0 & M_{12} & \cdots & M_{1m} \\ M_{21} & 0 & \cdots & M_{2m} \\ \vdots & \vdots & \ddots & \vdots \\ M_{m1} & M_{m2} & \cdots & 0 \end{bmatrix},$$

$$\mathbf{Z}_r = \begin{bmatrix} Z_{r1} & 0 & \cdots & 0 \\ 0 & Z_{r2} & \cdots & 0 \\ \vdots & \vdots & \ddots & \vdots \\ 0 & 0 & \cdots & Z_{rm} \end{bmatrix}, \text{ and}$$

$$Z_{ri} = R_{eqi} + R_{ri} + \left(\omega L_{ri} - \frac{1}{\omega C_{ri}} \right) j \ (i = 1, 2 \ldots, m).$$

Equation (3.22) can be simplified by omitting the term $\mathbf{i}_{r1}$, as illustrated below:

$$\frac{v_{t1}}{i_{t1}} = Z_{in1} = R_t + \left(\omega L_t - \frac{1}{\omega C_t} \right) j + \omega^2 \mathbf{M}^T \left(\omega \mathbf{M}_1 j + \mathbf{Z}_r \right)^{-1} \mathbf{M} \tag{3.23}$$

Here, Z_{in1} represents the fundamental input impedance of the SS-compensated SIMO-WPT system. The real and imaginary components of the term in Equation (3.23) can be separated as shown below:

$$\mathbf{Re}(Z_{\text{in1}}) = R_{\text{t}} + \omega^2 P(\mathbf{M}, \mathbf{M}_1, \mathbf{Z}_{\text{r}}, \omega) \tag{3.24}$$

$$\mathbf{Im}(Z_{\text{in1}}) = \left(\omega L_{\text{t}} - \frac{1}{\omega C_{\text{t}}} + \omega^2 Q(\mathbf{M}, \mathbf{M}_1, \mathbf{Z}_{\text{r}}, \omega) \right) j \tag{3.25}$$

Here, $P(\mathbf{M}, \mathbf{M}_1, \mathbf{Z}_{\text{r}}, \omega)$ and $Q(\mathbf{M}, \mathbf{M}_1, \mathbf{Z}_{\text{r}}, \omega)$ are scalars determined by the values of $\mathbf{M}$, $\mathbf{M}_1$, $\mathbf{Z}_{\text{r}}$, and ω. In theory, the values of $\mathbf{M}$, $\mathbf{M}_1$, and $\mathbf{Z}_{\text{r}}$ can be obtained by measuring the quantities v_{t1} and i_{t1} at different values of ω. In other words, Z_{in1} changes as the switching angular frequency ω is varied from the upper bound ω_{H} to the lower bound ω_{L}, or vice versa. Z_{in1} depends on the parameters L_{t}, C_{t}, R_{t}, $L_{\text{r}i}$, $C_{\text{r}i}$, $R_{\text{eq}i}$, $\mathbf{M}$, and $\mathbf{M}_1$.

$$\begin{cases} \dfrac{v_{\text{t1}l}}{i_{\text{t1}l}} = Z_{\text{in1}l} = f\left(L_{\text{t}}, C_{\text{t}}, R_{\text{t}}, L_{\text{r}i}, C_{\text{r}i}, R_{\text{eq}i}, \mathbf{M}, \mathbf{M}_1, \omega_l\right) \\ \omega_{\text{L}} \le \omega_l \le \omega_{\text{H}} \end{cases} \tag{3.26}$$

where n is the number of measured v_{t1} and i_{t1} for monitoring. By assuming that L_{t}, C_{t}, and R_{t} of the transmitter coil are preliminarily known in practice, the identification model for the parameters of the receiving resonators and mutual inductances can be derived as follows:

$$\min J = \mathbf{v}_{\text{t1est}} - \mathbf{v}_{\text{t1}} \tag{3.27}$$

$$\text{s.t. } \mathbf{v}_{\text{t1est}} = \mathbf{i}_{\text{t1}} \mathbf{Z}_{\text{in1}}$$

$$\omega_{\text{L}} \le \omega_l \le \omega_{\text{H}}$$

$$L_{\text{r}i\text{L}} \le L_{\text{r}i} \le L_{\text{r}i\text{H}}$$

$$C_{\text{r}i\text{L}} \le C_{\text{r}i} \le C_{\text{r}i\text{H}}$$

$$R_{\text{eq}i\text{L}} \le R_{\text{eq}i} \le R_{\text{eq}i\text{H}}$$

$$\mathbf{M}_{\text{L}} \le \mathbf{M} \le \mathbf{M}_{\text{H}}$$

$$\mathbf{M}_{1\text{L}} \le \mathbf{M}_1 \le \mathbf{M}_{1\text{H}}$$

where $\mathbf{Z}_{\text{in1}} = \begin{bmatrix} Z_{\text{in11}} & 0 & \cdots & 0 \\ 0 & Z_{\text{in12}} & \cdots & 0 \\ \vdots & \vdots & \ddots & \vdots \\ 0 & 0 & \cdots & Z_{\text{in1m}} \end{bmatrix}$. The estimated input voltages of the transmitter resonator $\mathbf{v}_{\text{t1est}} = [v_{\text{t1est1}}, v_{\text{t1est2}}, \ldots, v_{\text{t1est}n}]$ are calculated by $\mathbf{v}_{\text{t1est}} = \mathbf{i}_{\text{t1}}\mathbf{Z}_{\text{in1}}$ based on the measured transmitter currents $\mathbf{i}_{\text{t1}} = [i_{\text{t11}}, i_{\text{t12}}, \ldots, i_{\text{t1}n}]$ and the search parameters L_{ri}, C_{ri}, $R_{\text{eq}i}$, $\mathbf{M}$, and $\mathbf{M}_1$ at the switching angular frequencies ω_l ($l = 1, 2, \ldots, n$). The expressions of Z_{in1l} are provided in Equations (3.24) and (3.25). The objective of the identification model is to minimize the norm of the voltage differences between the estimated input voltages $\mathbf{v}_{\text{t1est}}$ and the measured input voltages $\mathbf{v}_{\text{t1}}$. The identified parameters L_{ri}, C_{ri}, $R_{\text{eq}i}$, $\mathbf{M}$, and $\mathbf{M}_1$ are searched within the lower bounds $L_{ri\text{L}}$, $C_{ri\text{L}}$, $R_{\text{eq}i\text{L}}$, $\mathbf{M}_{\text{L}}$, $\mathbf{M}_{\text{1L}}$ and the upper bounds $L_{ri\text{H}}$, $C_{ri\text{H}}$, $R_{\text{eq}i\text{H}}$, $\mathbf{M}_{\text{H}}$, $\mathbf{M}_{\text{1H}}$. The values of these bounds are determined empirically. The role of heuristic algorithms is to minimize the objective function defined in Equation (3.27). Compared to traditional optimization methods, heuristic algorithms can significantly enhance both the speed and accuracy of the identification process.

3.2.2 Critical Issue of the Equivalent Load Resistance

In practice, the identification accuracies are limited even when the system parameters are accurately determined, and measurement errors are minimized to a certain extent. The main reason for this limitation is that the relationship $R_{\text{L}} = \frac{\pi^2}{8} R_{\text{eq}}$ is not precise under certain operating conditions. As the load on the SS-compensated WPT system becomes lighter, the harmonic components become increasingly significant. Consequently, the amplitude of the total receiver current i_{r} deviates substantially from that of its fundamental component i_{r1}, as shown in Figure 3.13.

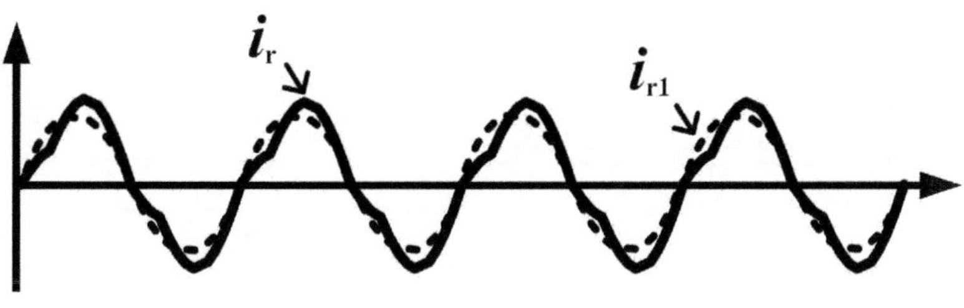

FIGURE 3.13
Schematic waveforms of a harmonic-distorted receiver current and its fundamental component.

As demonstrated in [1], the relationship $R_L = \frac{\pi^2}{8} R_{eq}$ is merely a simplified version of a more comprehensive and accurate formula:

$$R_L = \frac{\pi^2}{8(1+\kappa_h)} \left(R_{eq} - \frac{8 v_{eqD}}{\pi I_{r1}} \right) \tag{3.28}$$

$$\kappa_h = \frac{\cos\psi_3}{3}\left(\frac{I_{r3}}{I_{r1}}\right) + \frac{\cos\psi_5}{5}\left(\frac{I_{r5}}{I_{r1}}\right) + \cdots + \frac{\cos\psi_N}{N}\left(\frac{I_{rN}}{I_{r1}}\right) \tag{3.29}$$

$$v_{eqD} = \frac{v_{D1} + v_{D2} + v_{D3} + v_{D4}}{4} \tag{3.30}$$

where I_{r1}, I_{r3}, I_{r5}, and I_{rN} represent the amplitudes of the fundamental, third-order, fifth-order, and Nth-order harmonics of the receiver currents, respectively. The phase angles of these harmonics with respect to the fundamental current are denoted as $\psi_1, \psi_3, \psi_5, \ldots, \psi_N$. v_{eqD} represents the average value of the forward voltages across the four diodes in the diode-bridge rectifier. When the harmonics and diode forward voltages are ignored, the terms $\frac{I_{r3}}{I_{r1}}, \frac{I_{r5}}{I_{r1}}, \ldots, \frac{I_{rN}}{I_{r1}}$ and v_{eqD} become zero, leading to $\kappa_h=0$ and $v_{eqD}=0$. As a result, the formula in Equation (3.28) simplifies to the classic expression $R_L = \frac{\pi^2}{8} R_{eq}$.

However, the load model described in Equation (3.28) remains an intermediate representation. If i_r is highly distorted, the rectified current i_{rec} becomes discontinuous, as illustrated in Figure 3.14a. During the discontinuous intervals, the diode rectifier ceases to conduct, causing the receiving resonator to behave like an open circuit. As a result, the voltage v_r is mainly influenced by the transmitter-side currents, as depicted in Figure 3.15. It is important to note that the fundamental components of i_t dominate the response of the open-loop receiving resonator at the resonant frequency. This behavior can be further explained using Kirchhoff's voltage law.

$$v_r = M \frac{di_{t1}}{dt} \tag{3.31}$$

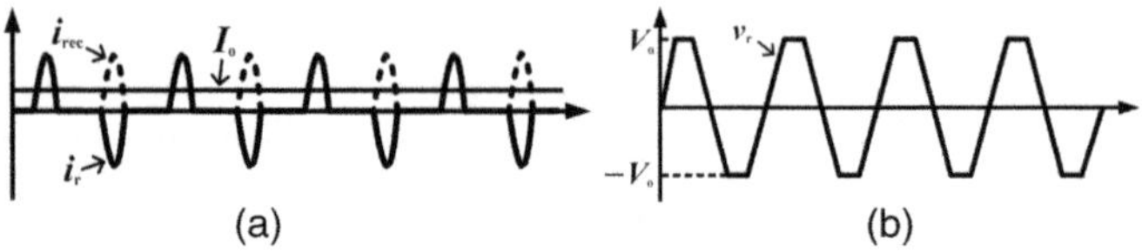

FIGURE 3.14
Schematic waveforms of (a) a highly distorted receiver current, its rectified form, and the corresponding output DC current, and (b) the output voltage of the receiving resonator.

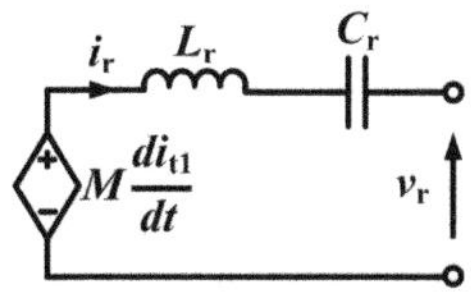

FIGURE 3.15
Equivalent circuit of the receiving resonator with a highly distorted receiver current.

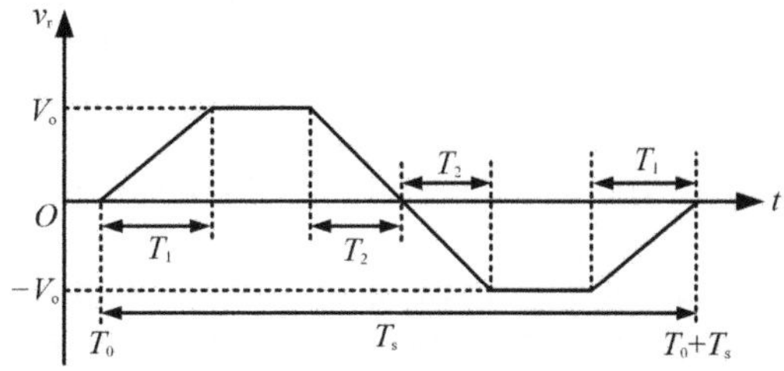

FIGURE 3.16
A detailed schematic of the receiving resonator's output voltage waveform over one cycle.

Clearly, the values of v_r are determined by both i_{t1} and M during the rising and falling edges of the trapezoidal waveform in Figure 3.14b. For further analysis, Figure 3.16 illustrates a schematic waveform of v_r over one cycle. Apparently, the durations of the edges (i.e., T_1 and T_2) are influenced by the total harmonic distortion (THD) of i_r.

Based on the schematic diagram of the trapezoidal waveform, a more comprehensive and accurate model can be derived as

$$R_L = \frac{\pi^2}{8(1+\kappa_h)G(\omega,T_1,T_2)}\left(R_{eq} - \frac{8v_{eqD}}{\pi I_{r1}}\right) \tag{3.32}$$

$$G(\omega,T_1,T_2) = \begin{cases} \frac{\sqrt{2}}{2}\sqrt{F_3(\omega T_1,\omega T_2)+\frac{F_2(\omega T_1)}{\omega T_1}+\frac{F_2(\omega T_2)}{\omega T_2}} & (T_1 \neq 0, T_2 \neq 0) \\ \frac{\sqrt{2}}{2}\sqrt{\frac{F_2(\omega T_1)}{\omega T_1}+F_1(\omega T_1)+\frac{1}{2}} & (T_1 \neq 0, T_2 = 0)\ (T_1 = 0, T_2 \neq 0) \\ \frac{\sqrt{2}}{2}\sqrt{\frac{F_2(\omega T_2)}{\omega T_2}+F_1(\omega T_2)+\frac{1}{2}} & \\ 1 & (T_1 = 0, T_2 = 0) \end{cases} \tag{3.33}$$

where $F_1(x) = \frac{\sin(x)}{x}$, $F_2(x) = \frac{1-\cos(x)}{x}$, and $F_3(x_1, x_2) = \frac{\cos(x_1)+\cos(x_2)-\cos(x_1+x_2)-1}{x_1 x_2}$. This advanced model represents a more comprehensive model of the nonlinear rectifier load. When the rectified current i_{rec} is continuous, T_1 and T_2 become zero and the trapezoidal waveform of v_r becomes a square-wave voltage. Under these conditions, $G(\omega, T_1, T_2)=1$, and the model in Equation (3.32) simplifies to the model in Equation (3.28).

From the analysis above, it is clear that the equivalent resistance of the nonlinear rectifier load in WPT systems is intricately complex. Its value is determined by the THD of the receiver current, which is closely tied to various parameters within the WPT system. If the equivalent load resistance (i.e., R_{eq}) is estimated using the heuristic algorithm outlined in Section 3.2.1, accurately estimating a wide range of output load resistances (i.e., R_L) requires the use of Equation (3.32). However, this equation relies on receiver-side current measurements to determine the parameters κ_h and $G(\omega, T_1, T_2)$, which is incompatible with the proposed transmitter-side control paradigm.

3.2.3 Auxiliary Circuit-Based Non-Resonant Frequency Method

To address the critical challenges posed by the nonlinear rectifier load, which can inaccurately affect the estimation of key WPT system parameters such as mutual inductances or coupling coefficients of the coupled coils, a non-resonant frequency method utilizing an auxiliary circuit was proposed in [5].

The fundamental concept of the auxiliary circuit-based non-resonant frequency method is to short-circuit the diode rectifier load. In practical applications, the short circuit can be achieved by adding a shunt switch to the DC side of the rectifier, as depicted in Figure 3.17a. The additional costs that are associated with incorporating this shunt switch are minimal. It can be implemented using a switch powered by the battery load. The power switch is

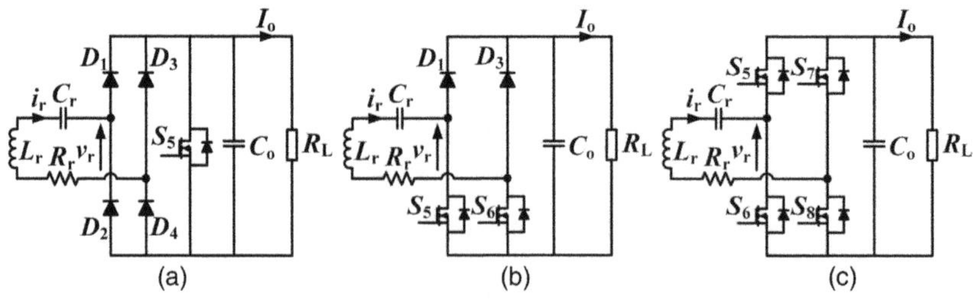

FIGURE 3.17
Equivalent circuits of WPT receivers with (a) a shunt power switch on the DC side, (b) a semi-bridgeless rectifier, and (c) a full-bridge rectifier.

activated upon receiving a simple message from the transmitter side, signaling the start of monitoring. This message can be embedded within the power signal, a method known as in-band communication, such as FSK or ASK. One of the in-band communication technologies we have developed will be discussed in the next chapter. If an active rectifier, such as the semi-bridgeless rectifier or the full-bridge rectifier depicted in Figure 3.17b and c, is employed for rectification, no extra hardware is required. The short circuit can be achieved by turning on the two switches on the lower bridge arms. The focus of the auxiliary circuit-based non-resonant frequency method discussed in this book is the coupling coefficient between the coupled coils. However, this method can be adapted to identify other parameters as well, provided the auxiliary circuit is appropriately designed.

By activating the shunt switch, the nonlinear rectifier load in the WPT system is effectively bypassed. The equivalent circuit of the SS-compensated WPT system in the frequency domain is depicted, as shown in Figure 3.18. Since the load is short-circuited, the equivalent load resistance R_{eq} can be approximated as close to zero. Given the high quality (Q) factors typically employed in WPT systems, harmonics are negligible under normal coupling conditions. Consequently, only the fundamental components need to be considered in the equivalent circuit. By substituting Equation (3.2) into Equation (3.1) to eliminate the term i_r and nullifying R_{eq}, the input impedance Z_{in} of the SS-compensated WPT system with the nonlinear rectifier load bypassed can be derived as

$$\mathbf{Re}(Z_{in}) = R_t + \frac{\omega^2 M^2}{R_r + \dfrac{\left(1 - \dfrac{\omega_{or}^2}{\omega^2}\right)^2 \omega^2 L_r^2}{R_r}} \tag{3.34}$$

$$\mathbf{Im}(Z_{in}) = \left(1 - \frac{\omega_{ot}^2}{\omega^2}\right)\omega L_t - \frac{\left(1 - \dfrac{\omega_{or}^2}{\omega^2}\right)\dfrac{\omega M^2}{L_r}}{\left(1 - \dfrac{\omega_{or}^2}{\omega^2}\right)^2 + \dfrac{R_r^2}{\omega^2 L_r^2}} \tag{3.35}$$

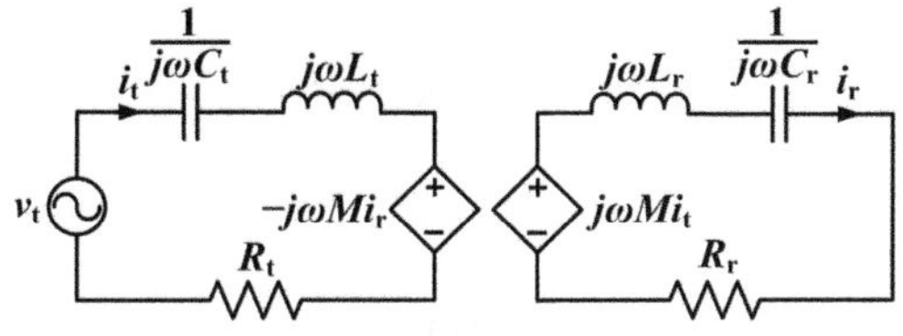

FIGURE 3.18
Equivalent circuit in the fundamental frequency domain of a typical SS-compensated WPT system with the nonlinear rectifier load bypassed.

where $\omega_{ot}^2 = \frac{1}{L_t C_t}$ and $\omega_{or}^2 = \frac{1}{L_r C_r}$ are the resonant angular frequencies of the transmitting and receiving resonators, respectively. Typically, the term $\frac{R_r^2}{\omega^2 L_r^2}$ is negligible due to the high Q factors of receiving coils. As a result, Equation (3.35) can be simplified as

$$\mathbf{Im}(Z_{in}) = \begin{cases} \omega L_t \left[\left(1 - \frac{\omega_{ot}^2}{\omega^2}\right) - \frac{k^2}{1 - \frac{\omega_{or}^2}{\omega^2}} \right] (\omega \neq \omega_{or}) \\ \left(1 - \frac{\omega_{ot}^2}{\omega_{or}^2}\right) \omega_{or} L_t \; (\omega = \omega_{or}) \end{cases} \tag{3.36}$$

where k is the coupling coefficient of the coupled coils that satisfy $k = \frac{M}{\sqrt{L_t L_r}}$ (where $0 \leq k \leq 1$). The equivalent reactance $\mathbf{Im}(Z_{in})$ can exhibit inductive, capacitive, or zero characteristics in Equation (3.36). Based on Equation (3.36), the following conditions apply:

a. If $\omega = \omega_{or} > \omega_{ot}$, $\mathbf{Im}(Z_{in})$ is inductive for any k.
b. If $\omega = \omega_{or} < \omega_{ot}$, $\mathbf{Im}(Z_{in})$ is capacitive for any k.
c. If $\omega = \omega_{or} = \omega_{ot}$, $\mathbf{Im}(Z_{in})$ is zero for any k.

By assuming $k_c = \sqrt{\left(1 - \frac{\omega_{ot}^2}{\omega^2}\right)\left(1 - \frac{\omega_{or}^2}{\omega^2}\right)}$, $\mathbf{Im}(Z_{in})$ is inductive if

$$\begin{cases} k^2 < k_c^2 \; (\omega > \omega_{or}) \\ k^2 > k_c^2 \; (\omega < \omega_{or}) \end{cases} \tag{3.37}$$

$\mathbf{Im}(Z_{in})$ is capacitive if

$$\begin{cases} k^2 < k_c^2 \; (\omega < \omega_{or}) \\ k^2 > k_c^2 \; (\omega > \omega_{or}) \end{cases} \tag{3.38}$$

$\mathbf{Im}(Z_{in})$ is zero if

$$k^2 = k_c^2 \; (\omega \neq \omega_{or}) \tag{3.39}$$

The reactance characteristics of $\mathbf{Im}(Z_{in})$ are summarized in Table 3.2, where the coupling coefficient k ranges from 0 to 1. ω_o represents the resonant

TABLE 3.2

Characteristics of the Input Reactance Im(Z_{in})

$\omega_{ot} < \omega_{or}$				
	$\omega_{ot} < \omega_{or} < \omega$	$\omega_{ot} < \omega < \omega_{or}$	$\omega < \omega_{ot} < \omega_{or}$	$\omega_{ot} < \omega_{or} = \omega$
$k < k_c$	Inductive	Unavailable	Capacitive	Inductive
$k > k_c$	Capacitive	Capacitive	Inductive	Inductive
$k = k_c$	Zero	Unavailable	Zero	Inductive
$\omega_{ot} > \omega_{or}$				
	$\omega_{or} < \omega_{ot} < \omega$	$\omega_{or} < \omega < \omega_{ot}$	$\omega < \omega_{or} < \omega_{ot}$	$\omega = \omega_{or} < \omega_{ot}$
$k < k_c$	Inductive	Unavailable	Capacitive	Capacitive
$k > k_c$	Capacitive	Capacitive	Inductive	Capacitive
$k = k_c$	Zero	Unavailable	Zero	Capacitive

$\omega_{ot} = \omega_{or} = \omega_o$			
	$\omega < \omega_o$	$\omega > \omega_o$	$\omega = \omega_o$
$k < k_c$	Capacitive $(0 < k \le \frac{\omega_o^2}{\omega^2} - 1)$	Inductive $(0 < k \le 1 - \frac{\omega_o^2}{\omega^2})$	Zero
$k > k_c$	Inductive $(\frac{\omega_o^2}{\omega^2} - 1 < k \le 1)$	Capacitive $(1 - \frac{\omega_o^2}{\omega^2} < k \le 1)$	Zero
$k = k_c$	Zero	Zero	Zero

frequency of both transmitting and receiving resonators when their resonant frequencies are identical. Using the reactance characteristics of **Im**(Z_{in}) outlined in Table 3.2, the behaviors of the equivalent impedance Z_{in} in the k–ω plane are illustrated in Figure 3.19 for the three possible cases of the typical SS-compensated WPT system with the nonlinear rectifier load bypassed.

When the operating angular frequency ω and the coupling coefficient k of the WPT system lie within the inductive or capacitive zones (including the inductive or capacitive lines), the equivalent input impedance Z_{in} exhibits inductive or capacitive characteristics, respectively. When ω and k are located on the zero lines, Z_{in} becomes purely resistive. The intersection points of any k with the zero lines define the point A and the point B for any compensation scenario (i.e., $\omega_{ot}<\omega_{or}$, $\omega_{ot}>\omega_{or}$, or $\omega_{ot}=\omega_{or}=\omega_o$). Point A is determined by sweeping ω from the minimum angular frequency ω_{min} to the maximum angular frequency ω_{max}, while point B is determined by sweeping ω from ω_{max} to ω_{min}. Based on Equation (3.39), the analytical expressions for the zero lines passing through the point A and the point B are as follows:

$$k = \sqrt{\frac{\omega_{ot}^2}{\omega^2} - 1} \cdot \sqrt{\frac{\omega_{or}^2}{\omega^2} - 1} \qquad (3.40)$$

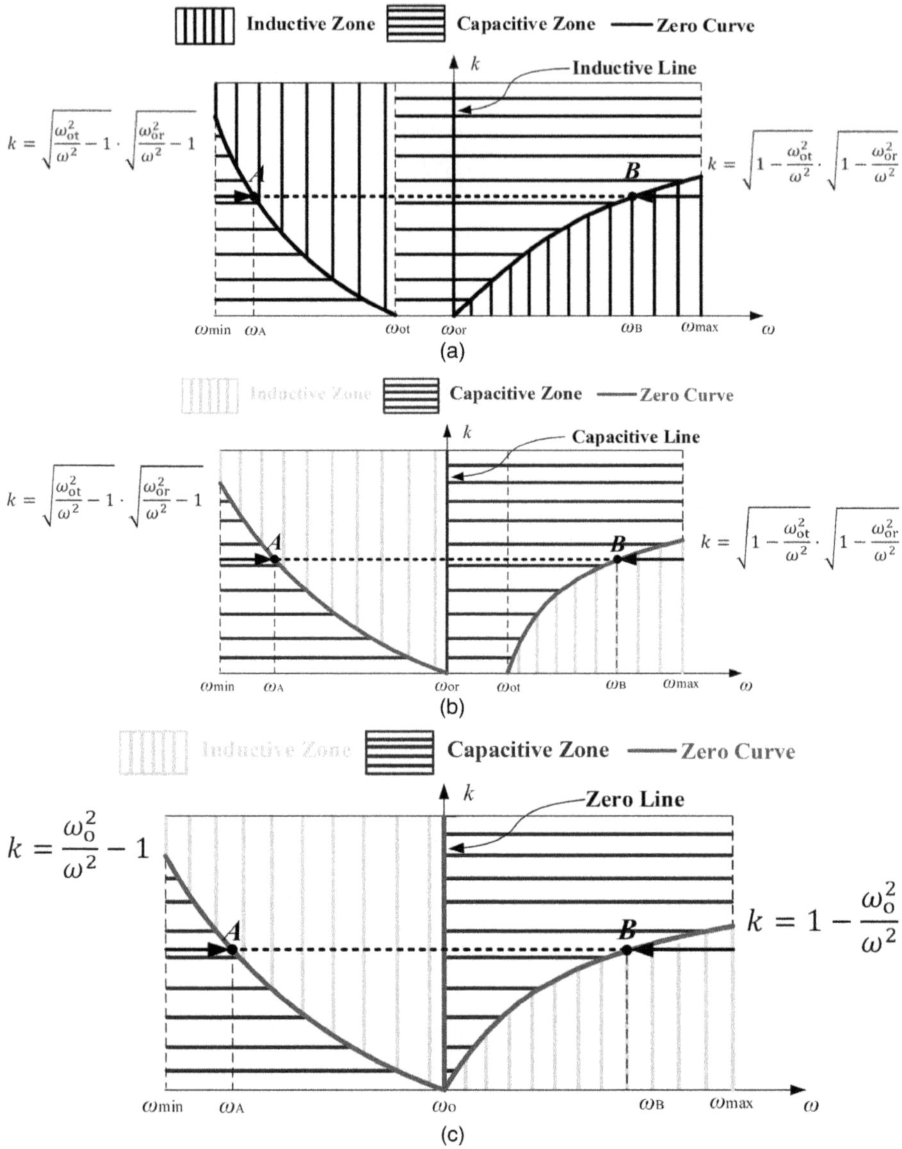

FIGURE 3.19
Characteristics of the input equivalent impedance Z_{in} in the k-ω plane: (a) $\omega_{ot}<\omega_{or}$, (b) $\omega_{ot}>\omega_{or}$, and (c) $\omega_{ot}=\omega_{or}=\omega_o$. (Adapted from reference [5].)

$$k=\sqrt{1-\frac{\omega_{ot}^2}{\omega^2}}\cdot\sqrt{1-\frac{\omega_{or}^2}{\omega^2}} \tag{3.41}$$

The angular frequencies corresponding to point A and the point B are ω_A and ω_B, respectively. Using Equations (3.40) and (3.41), the coupling coefficient of the system can be estimated using

$$k=\sqrt{\frac{\omega_{ot}^2}{\omega_A^2}-1}\cdot\sqrt{\frac{\omega_{or}^2}{\omega_A^2}-1}\ (\omega_A<\omega_{or}) \tag{3.42}$$

or

$$k=\sqrt{1-\frac{\omega_{ot}^2}{\omega_B^2}}\cdot\sqrt{1-\frac{\omega_{or}^2}{\omega_B^2}}\ (\omega_B>\omega_{or}) \tag{3.43}$$

It is preferable to use Equation (3.43) rather than Equation (3.42) to estimate the coupling coefficient in practice, as Z_{in} operates in the inductive region, which facilitates zero-voltage switching (ZVS) for the transmitter-side inverter. Moreover, C_t and C_r are typically designed to compensate L_t and L_r, respectively, ensuring the system operates at the desired resonant angular frequency ω_o for zero phase angle (ZPA). Achieving ZPA minimizes the apparent power rating of the power supply while maximizing power transfer efficiency. When $\omega_{ot}=\omega_{or}=\omega_o$, the equations for estimating the coupling coefficient can be simplified as

$$k=\frac{\omega_o^2}{\omega_A^2}-1\ (\omega_A<\omega_o) \tag{3.44}$$

or

$$k=1-\frac{\omega_o^2}{\omega_B^2}\ (\omega_B>\omega_o) \tag{3.45}$$

Apparently, once the critical frequencies ω_A and ω_B that achieve ZPA are identified, the coupling coefficient of the WPT system can be readily determined. In practice, transmitter current (i.e., i_t) is sensed and input into a zero-crossing comparator, as exhibited in Figure 3.20. The comparator output (i.e., i_{tsyn}) and the switching signal S_1 are fed into an XOR logic gate (or an equivalent logical circuit) to generate the signal S_{XOR}. Here, the switching signal S_1 corresponds to the input voltage of the transmitting resonator (i.e., v_t). When i_t is positive, i_{tsyn} generates a positive pulse. If S_1 and i_{tsyn} are in phase, it indicates that v_t and i_t are synchronized. The S_{XOR} signal reflects the phase difference between S_1 and i_{tsyn}. As the pulse width of S_{XOR} narrows, the phase shift between S_1 and i_{tsyn} decreases. When the pulse width

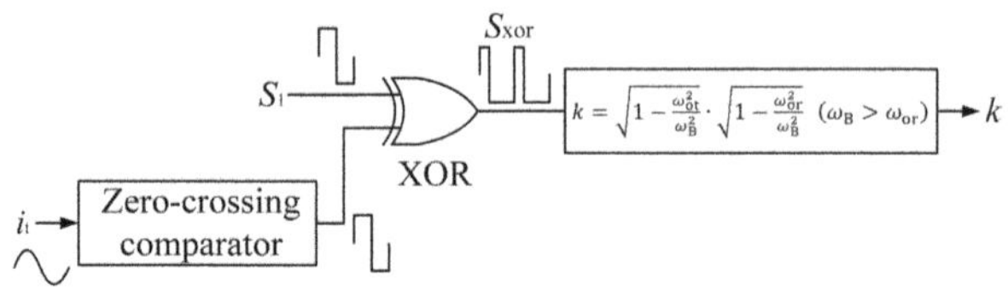

FIGURE 3.20
Block diagram illustrating the practical implementation of the monitoring strategy with auxiliary circuits.

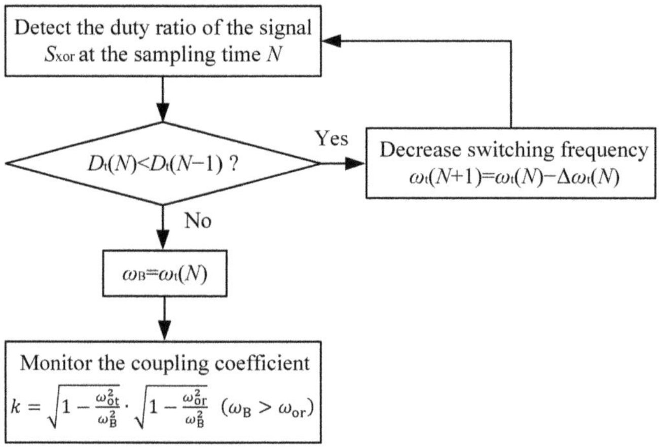

FIGURE 3.21
Flowchart of the monitoring strategy using auxiliary circuits to determine the critical angular frequency ω_B and calculate the coupling coefficient k. (Adapted from reference [5].)

of S_{XOR} reduces to zero, the system reaches either point A (for ω_A) or point B (for ω_B). As we have noted, point B is a more advantageous option compared to point A.

To determine ω_B, the operating frequency of the inverter is scanned starting from the upper bound angular frequency ω_{max}. Figure 3.21 illustrates the flowchart for identifying ω_B and subsequently calculating k. The duty cycle of the signal S_{XOR} is measured and digitized as $D_t(N)$. If the detected duty cycle is lower than the previous value (i.e., $D_t(N) < D_t(N-1)$), the switching frequency will be reduced by $\Delta\omega_t(N)$. This iterative process continues until D_tx(N) approaches zero, indicating that ω_B has been reached. Once ω_B is determined, the coupling coefficient k can be derived using Equation (3.43) or (3.45). The schematic waveforms of the main signals are depicted in Figure 3.22.

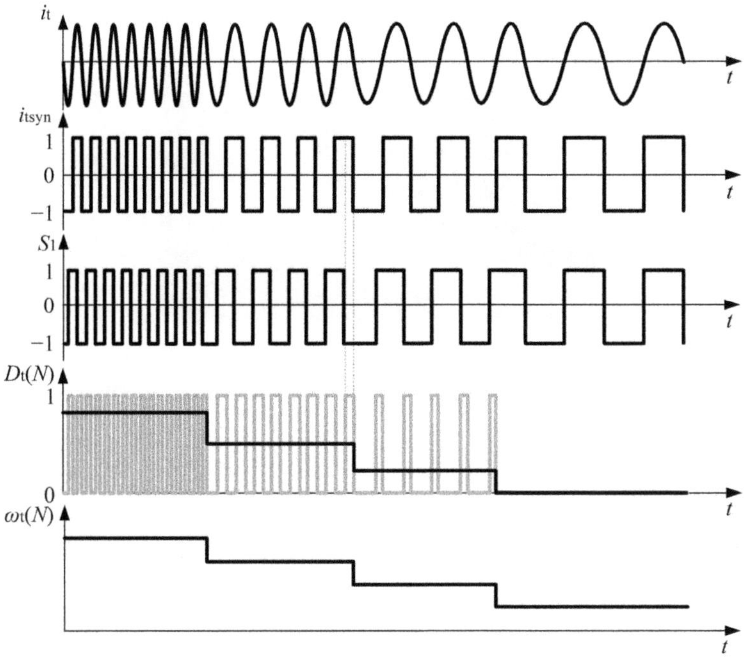

FIGURE 3.22
Schematic waveforms of the main signals for the monitoring strategy with auxiliary circuits. (Adapted from reference [5].)

3.3 Experimental Verification

To validate the analysis presented in Section 3.2, experiments are conducted on WPT systems with different parameters.

3.3.1 Heuristic Algorithm-Based Method

Experiments are conducted on an SS-compensated WPT system, as shown in Figure 3.23. Both the transmitting and receiving coils are 11 turns with the diameter of 31 cm. The main electrical parameters of the WPT system are listed in Table 3.3. The switching frequency is swept from 90 to 110 kHz in increments of 1 kHz. The distance between the coils is varied from 10 to 20 cm in steps of 2 cm. The corresponding mutual inductances are 15.515, 12.333, 10.194, 8.431, 7.021, and 5.935 μH, respectively. The fundamental and peak values of v_t and i_t for the WPT system are derived from the waveforms captured in MATLAB via National Instruments' VISA using the oscilloscope. Figure 3.24 compares the waveforms directly captured from the oscilloscope with those obtained in MATLAB for the WPT system, operating with

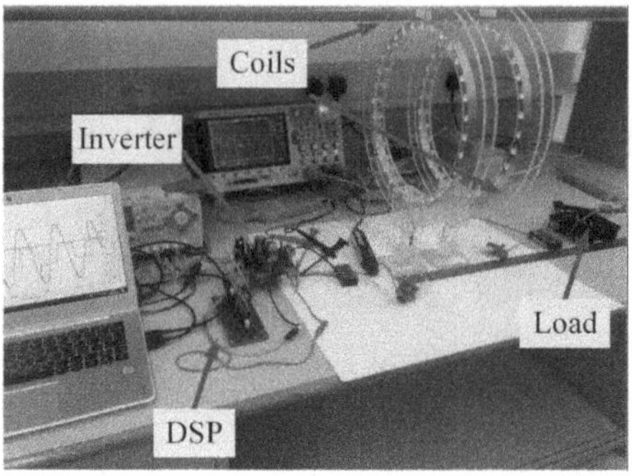

FIGURE 3.23
Photograph of the experimental setup for validating the heuristic algorithm-based method. (Adapted from reference [4].)

TABLE 3.3
Main Electrical Parameters of the SS-Compensated WPT System for Validating the Heuristic Algorithm-Based Method

Description	Symbol	Value
Inductance of the transmitting coil	L_t	91.78 μH
Compensated capacitance of the transmitter	C_t	27.62 nF
ESR of the transmitting resonator	R_t	0.7 Ω
Inductance of the receiving coil	L_r	92.05 μH
Compensated capacitance of the receiver	C_r	27.2 nF
ESR of the receiving resonator	R_r	1.01 Ω
Input DC voltage	V_{DC}	5 V
Nominal resonant frequency	f_o	100 kHz

a load resistance of 10 Ω and a coil distance of 10 cm, at switching frequencies of 90 kHz, 100 kHz (resonant frequency), and 110 kHz. Apparently, the waveforms obtained in MATLAB closely match those captured by the oscilloscope. Therefore, we can confidently conclude that the processed fundamental and peak signals (such as v_{t1}, i_{t1}, V_{t1}, and I_{t1}) are accurately measured.

Using the obtained data, an adaptive differential evolution (ADE) algorithm (renowned for its exceptional global optimization performance) is employed to identify the mutual inductance between the coupled coils and the key parameters of the receivers. The main parameters of the implemented ADE algorithm are given in Table 3.4. The population size of the ADE algorithm is tuned to be P_{size}= 7×nvar, where nvar represents the number of monitored variables. The maximum generation $\max_{gen}$ is tuned to implement fast convergence.

Twelve operating cases are evaluated: (i) load resistance of 10 Ω and coil distance of 10 cm; (ii) load resistance of 10 Ω and coil distance of 12 cm; (iii) load

(a)

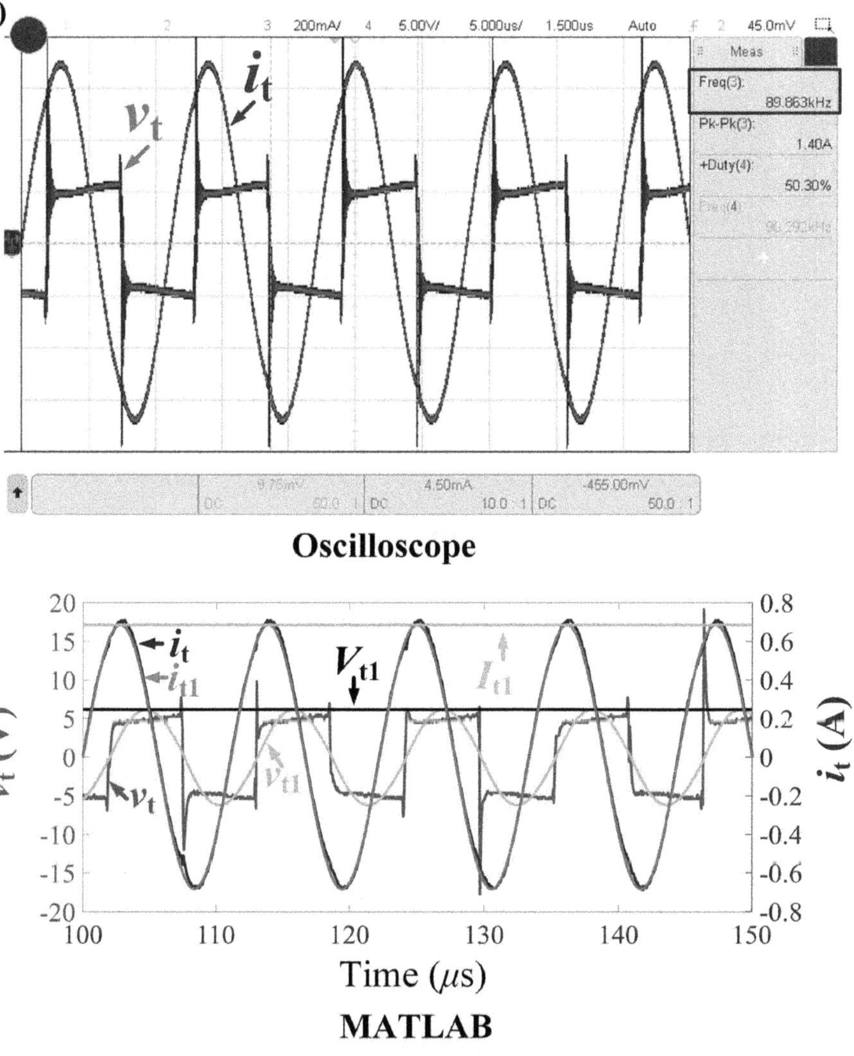

FIGURE 3.24
Waveforms of the input AC voltage (v_t) and transmitter current (i_t) directly captured from the oscilloscope and captured in MATLAB for the WPT system with a load resistance of 10 Ω and a coil distance of 10 cm at switching frequencies of (a) 90 kHz, (b) 100 kHz, and (c) 110 kHz.

(Continued)

(b)

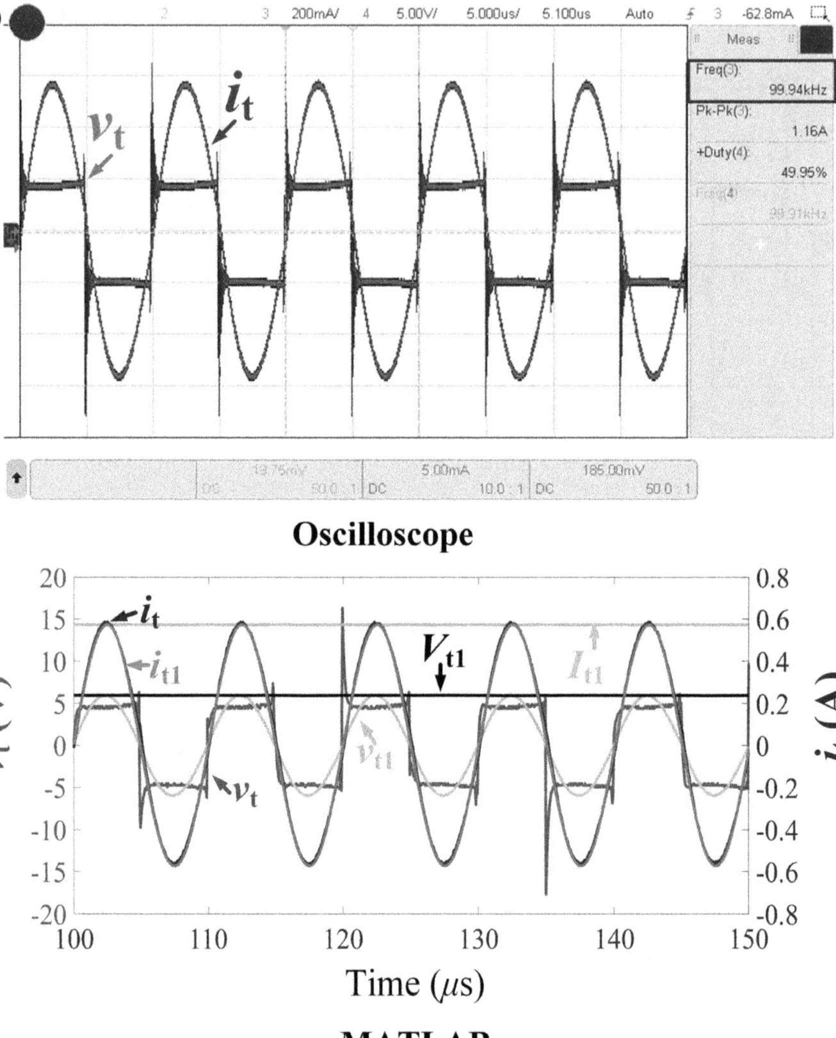

FIGURE 3.24 (*Continued*)
Waveforms of the input AC voltage (v_t) and transmitter current (i_t) directly captured from the oscilloscope and captured in MATLAB for the WPT system with a load resistance of 10 Ω and a coil distance of 10 cm at switching frequencies of (a) 90 kHz, (b) 100 kHz, and (c) 110 kHz.

(*Continued*)

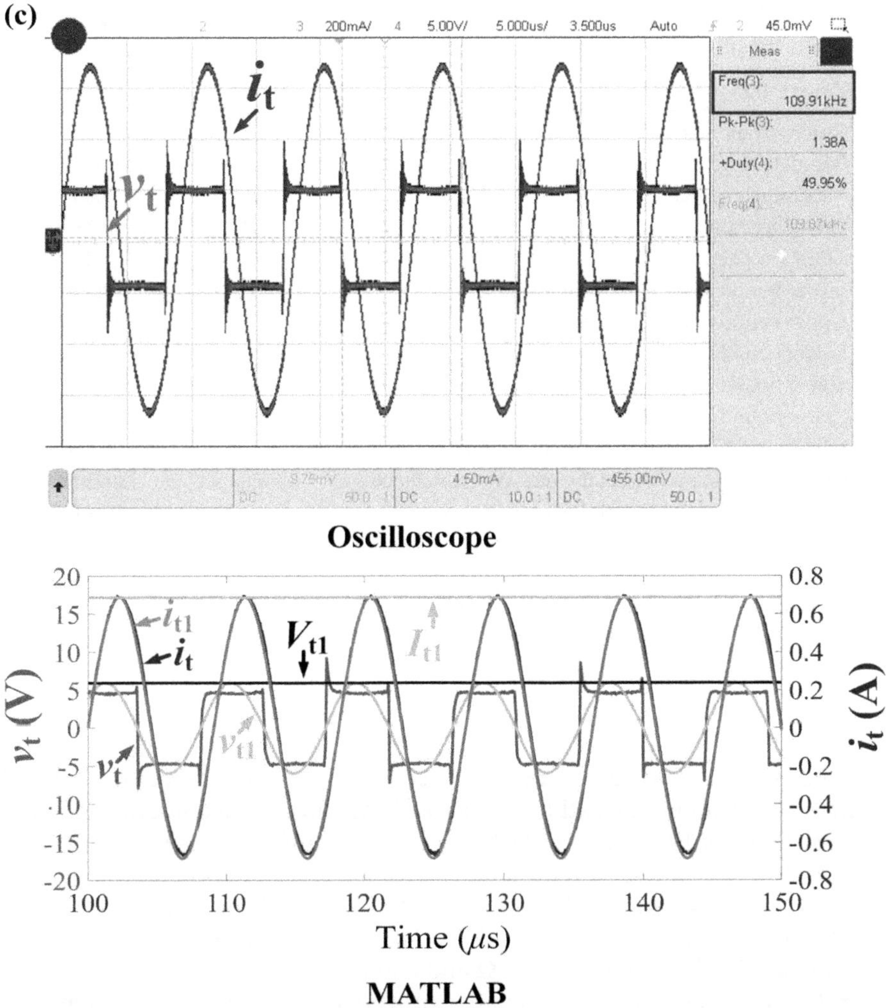

FIGURE 3.24 (*Continued*)
Waveforms of the input AC voltage (v_t) and transmitter current (i_t) directly captured from the oscilloscope and captured in MATLAB for the WPT system with a load resistance of 10 Ω and a coil distance of 10 cm at switching frequencies of (a) 90 kHz, (b) 100 kHz, and (c) 110 kHz.

TABLE 3.4
Main Parameters of the ADE Algorithm

Description	Symbol	Value
Chromosome size	C_{size}	10
Population size	P_{size}	64
Maximum generations	max_{gen}	2000
Crossover rate	P_c	0.26
Mutation rate	P_m	0.2
Lower bound of the differential weight	F_{low}	0.1
Upper bound of the differential weight	F_{up}	0.8
Lower bound of the crossover rate	P_{low}	0.2
Upper bound of the crossover rate	P_{up}	0.6
Lower bound of the differential weight	F_{low}	0.1
Lower searching bound of the receiver inductance	L_{rL}	80 μH
Upper searching bound of the receiver inductance	L_{rH}	100 μH
Lower searching bound of the receiver capacitance	C_{rL}	20 nF
Upper searching bound of the receiver capacitance	C_{rH}	50 nF
Lower searching bound of the equivalent load	R_{eqL}	0 Ω
Upper searching bound of the equivalent load	R_{eqH}	12 Ω
Lower searching bound of the mutual inductance	M_L	5 μH
Upper searching bound of the mutual inductance	M_H	20 μH

resistance of 10 Ω and coil distance of 14 cm; (iv) load resistance of 10 Ω and coil distance of 16 cm; (v) load resistance of 10 Ω and coil distance of 18 cm; (vi) load resistance of 10 Ω and coil distance of 20 cm; (vii) load resistance of 4.7 Ω and coil distance of 10 cm; (viii) load resistance of 4.7 Ω and coil distance of 12 cm; (ix) load resistance of 4.7 Ω and coil distance of 14 cm; (x) load resistance of 4.7 Ω and coil distance of 16 cm; (xi) load resistance of 4.7 Ω and coil distance of 18 cm; (xii) load resistance of 4.7 Ω and coil distance of 20 cm.

For the first case with load resistance of 10 Ω and coil distance of 10 cm, the ADE algorithm converges at 233 generations, the parameters L_r, C_r, R_{eq}, and M can be consistently monitored for generations 233, 500, 1000, and 2000. After convergence, the relative errors for L_r, C_r, and M are less than 0.5%, while those for R_{eq} are below 1%. The fitness values for the remaining 11 cases are 240, 200, 90, 250, 240, 250, 160, 200, 350, 230, and 360, respectively, further validating the ADE's capability to identify global optimal points. The average relative errors of the identified parameters L_r, C_r, R_{eq}, and M across all 12 cases are below 1.5%, 1.5%, 3%, and 1%, respectively. This verifies the high accuracy of the ADE algorithm in identifying the key parameters of receivers in WPT systems. To demonstrate the superior global optimization capability of the ADE algorithm over the conventional GA, experiments are conducted on the WPT system across 12 operating cases. Compared to GA, the ADE algorithm reduces estimation errors by approximately 1% for L_r,

0.8% for C_r, 10.4% for R_{eq}, and 4.6% for M. Besides, the standard deviations of the monitored parameters across all 12 cases are significantly reduced, with reductions of approximately 42.01% for L_r, 40.75% for C_r, 78.84% for $R_{eq}=10\ \Omega$, 75.6% for $R_{eq}=4.7\ \Omega$, 83.42% for $D=10$ cm, 88.26% for $D=12$ cm, 78.87% for $D=14$ cm, 96.83% for $D=16$ cm, 90.91% for $D=18$ cm, and 98.35% for $D=20$ cm. This highlights that a heuristic algorithm with superior global optimization capabilities is preferable for parameter identification, as it provides more accurate and stable estimations.

3.3.2 Nonlinear Rectifier Load

Experiments are conducted on a prototype with the main parameters given in Table 3.5. The rated power is 25 W. The RIGOL DP811A and DP832 DC supply power to the main WPT system and the auxiliary circuits, respectively. The full-bridge inverter is designed with the silicon carbide (SiC) Metal-Oxide-Semiconductor Field-Effect Transistor (MOSFET) G3R75MT12D, while an FY6300-60M function generator supplies an 85 kHz complementary switching signal to drive the inverter. The compensated capacitors, along with the filter capacitor, are constructed using metallized polyester film capacitors. The diode-bridge rectifier is composed of ISL9R860P2 diodes, which exhibit forward voltages of approximately 0.7 V (measured at 25°C and 1 A forward current) and junction capacitances of around 30 pF (at 85 kHz). Therefore, the equivalent diode forward voltage v_{eqD} can be calculated from Equation (3.30) as 0.7 V. The load is implemented using a variable resistor, specifically the TERCO-MV1100 model. The load resistances are manually adjusted to 10, 35, and 285 Ω. To ensure measurement accuracy, average steady-state values are utilized. The waveforms of the harmonic components of i_r are reconstructed in MATLAB using the exported measured data from the oscilloscope.

The waveforms of v_r, i_r, i_{rec}, and I_o under $R_L=10\ \Omega$ (heavy load condition) are shown in Figure 3.25a. As observed, v_r displays a rectangular waveform, while i_{rec} remains continuous. The fast Fourier transform (FFT) spectra of v_r and i_r are presented in Figure 3.25b and c, respectively. From the FFT spectra,

TABLE 3.5

Main Electrical Parameters of the SS-Compensated WPT System for Validating the Equivalent Load Equations

Parameters	Value	Parameters	Value
Transmitter inductance L_t	49.1 μH	Receiver inductance L_r	48.9 μH
Transmitter capacitance C_t	70 nF	Receiver capacitance C_r	71.5 nF
Transmitter resistance R_t	0.13 Ω	Receiver resistance R_r	0.12 Ω
Resonant frequency f_o	85 kHz	Mutual inductance M	15.05 μH
Input DC voltage V_{DC}	12 V	Output DC capacitance C_o	20 μF
Diode forward voltage V_D	0.7 V		

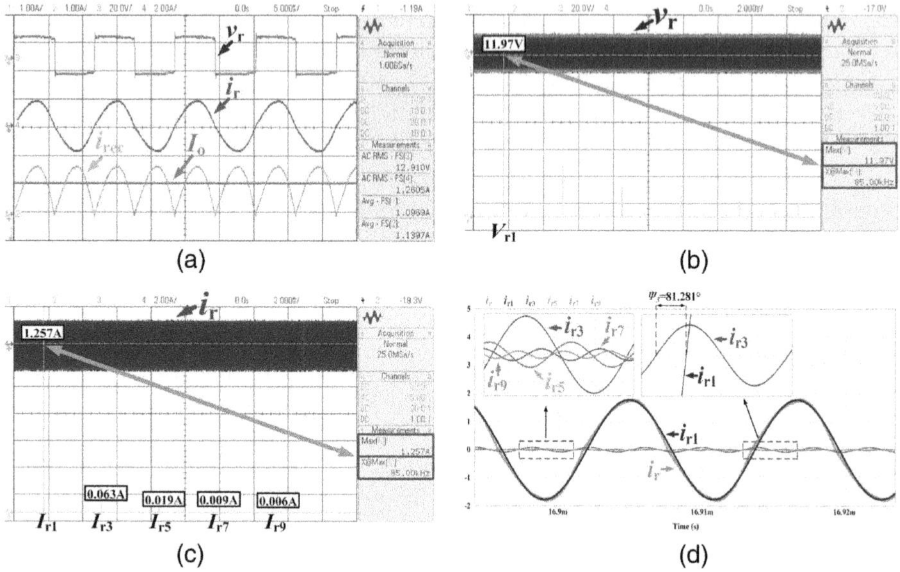

FIGURE 3.25
Waveforms and FFT spectra of major receiver measurements for the WPT system with $R_L = 10\,\Omega$ (heavy load condition): (a) waveforms of v_r, i_r, i_{rec}, and I_o, (b) FFT spectra of v_r, (c) FFT spectra of i_r, and (d) waveforms of i_{r1}, i_{r3}, i_{r5}, i_{r7}, and i_{r9}. (Adapted from reference [1].)

V_{r1} and I_{r1} are measured as 11.97 V and 1.257 A, respectively. Accordingly, the equivalent resistance R_{eq} is derived as 9.523 Ω. When the conventional equation $R_L = \frac{\pi^2}{8} R_{eq}$ is used, R_L is determined to be 11.75 Ω, leading to a 17.49% error. However, by counting for the diode forward voltages, the conventional equation $R_L = \frac{\pi^2}{8} R_{eq}$ is modified to the equation $R_L = \frac{\pi^2}{8}\left(R_{eq} - \frac{8 v_{eqD}}{\pi I_{r1}}\right)$. This adjustment yields an R_L value of 9.999 Ω, reducing the error to just 0.01%. Therefore, the inclusion of the augmented terms representing the diode forward voltages helps improve the accuracy of the nonlinear rectifier load model in WPT systems. For WPT systems operating under heavy load conditions, the advanced Equations (3.28) and (3.30) may not be necessary, as the equation $R_L = \frac{\pi^2}{8}\left(R_{eq} - \frac{8 v_{eqD}}{\pi I_{r1}}\right)$ provides sufficient accuracy.

Figure 3.26 presents the corresponding waveforms and FFT spectra for the WPT system with $R_L = 35\,\Omega$ (light load condition). The voltage v_t remains square-shaped, while the rectifier current i_{rec} is continuous. However, the resonant current i_r exhibits distortion due to harmonic components. Based on the FFT spectra, V_{r1} and I_{r1} are measured as 30.1 V and $I_{s1} = 1.027$ A, respectively. Consequently, R_{eq} can be calculated as 29.309 Ω. Based on the conventional equation $R_L = \frac{\pi^2}{8} R_{eq}$, R_L is calculated to be 36.159 Ω, resulting in a 3.31% error.

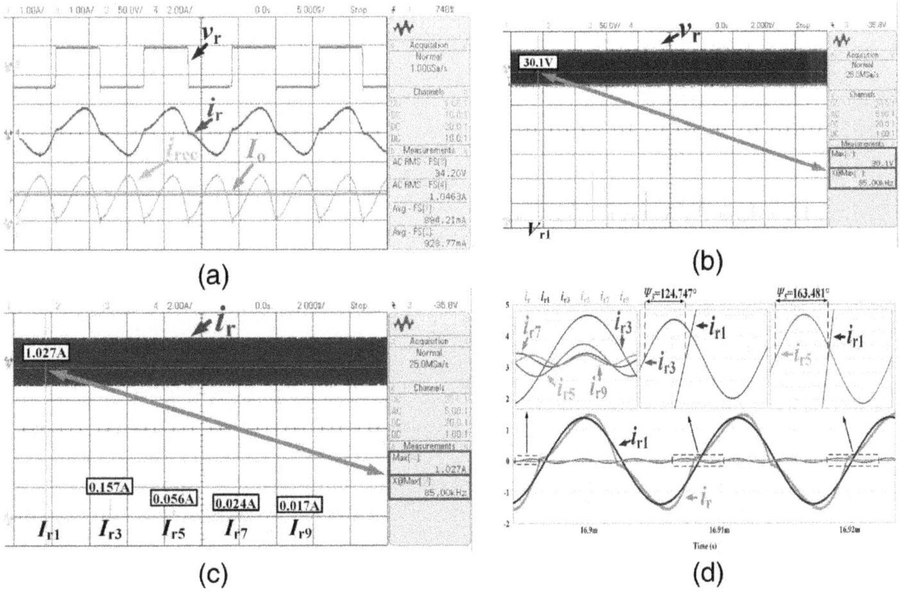

FIGURE 3.26
Waveforms and FFT spectra of major receiver measurements for the WPT system with $R_L=35\,\Omega$ (light load condition): (a) waveforms of v_r, i_r, i_{rec}, and I_o, (b) FFT spectra of v_r, (c) FFT spectra of i_r, and (d) waveforms of i_{r1}, i_{r3}, i_{r5}, i_{r7}, and i_{r9}. (Adapted from reference [1].)

By using the modified equation $R_L = \dfrac{\pi^2}{8}\left(R_{eq} - \dfrac{8v_{eqD}}{\pi I_{r1}}\right)$, R_L is calculated to be 34.017 Ω, yielding a reduced error of 2.81%. I_{r3} and I_{s5} are measured to be 0.157 and 0.056 A based on the FFT spectra in Figure 26c. Additionally, ψ_3 and ψ_5 are measured as 124.747° and 163.481° from the waveforms in Figure 26d. Using these values, the coefficient k_h is calculated as –0.03941 for $N=5$ (harmonics of orders higher than 5 are negligible). Based on Equation (3.28), R_L is calculated to be 35.413 Ω, with a corresponding error of 1.18%. Compared to 2.81%, it is reduced by more than 1.5%, validating the analysis that the intermediate model based on Equation (3.28) provides greater accuracy than the conventional model for WPT systems with light load conditions. For the most comprehensive model in Equation (3.32), since its coefficient $G(\omega, T_1, T_2)$ of the square waveform v_r is 1 (i.e., $T_1=T_2=0$), the calculated load error should match that of the model in Equation (3.28).

The waveforms and FFT spectra for the WPT system with $R_L=285\,\Omega$ (ultra-light load condition) are depicted in Figure 3.27. Here, v_r exhibits a trapezoidal waveform with high-frequency oscillations, while i_{rec} is discontinuous. The current i_r is highly distorted due to harmonics, and both i_r and i_{rec} display high-frequency oscillations at zero-current intervals. From the spectra, V_{r1} and I_{r1} are measured to be 83.6 V and 0.454 A, respectively. Consequently,

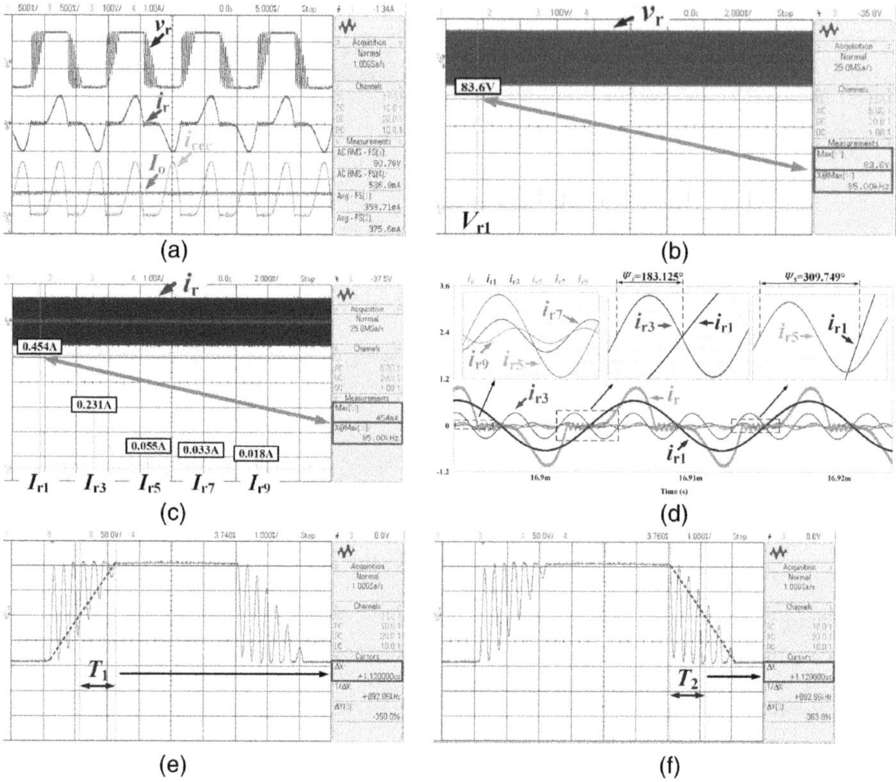

FIGURE 3.27
Waveforms and FFT spectra of major receiver measurements for the WPT system with $R_L = 285\,\Omega$ (ultralight load condition): (a) waveforms of v_r, i_r, i_{rec}, and I_o, (b) FFT spectra of v_r, (c) FFT spectra of i_r, (d) waveforms of i_{r1}, i_{r3}, i_{r5}, i_{r7}, and i_{r9}, (e) measurements of T_1 at the rising edge of v_r, and (f) measurements of T_2 at the falling edge of v_r. (Adapted from reference [1].)

R_{eq} is calculated to be 184.141 Ω. The corresponding R_L is calculated as 227.175 Ω using the conventional equation $R_L = \frac{\pi^2}{8} R_{eq}$ and 222.331 Ω using the modified equation $R_L = \frac{\pi^2}{8}\left(R_{eq} - \frac{8 v_{eqD}}{\pi I_{r1}}\right)$. The associated errors are 20.29% and 21.99%, respectively. Obviously, the conventional model even with the augmented terms accounting for diode forward voltages is substantially flawed. I_{r3} and I_{r5} are measured to be 0.231 and 0.055 A based on the FFT spectra in Figure 27c. Additionally, ψ_3 and ψ_5 are measured as 183.125° and 309.749° from the waveforms in Figure 27d. Using these values, the coefficient k_h is calculated as −0.15383 for $N = 5$ (harmonics of orders higher than 5 are negligible). Based on Equation (3.28), R_L is determined to be 262.75 Ω, with a corresponding error of 7.81%. Compared to the conventional model, this intermediate model offers improved accuracy. However, its precision remains insufficient for certain practical applications, as the error still exceeds 5%. The time periods of the high-frequency oscillations (i.e., T_1 and T_2) are directly measured from the waveforms in Figure 3.27e and f. To ensure precise measurement, virtual rising and falling edges are drawn by connecting the steady-state voltage levels. In this case, $T_1 = T_2 = 1.12$ μs. Consequently, the coefficient $G(\omega, T_1, T_2)$ is calculated as 0.9414 using Equation (3.33). Applying Equation (3.32), R_L is estimated as 279.11 Ω, with a corresponding relative error of 2.07%. Evidently, the model based on Equation (3.32) is the only reliable one for WPT systems under ultralight load conditions. It represents the most comprehensive and general model we have developed.

Waveforms and FFT spectra of major receiver measurements for the WPT system with $R_L = 285\,\Omega$ (ultralight load condition): (i) waveforms of v_r, i_r, i_{rec}, and I_o, (ii) FFT spectra of v_r, (iii) FFT spectra of i_r, (iv) waveforms of i_{r1}, i_{r3}, i_{r5}, i_{r7}, and i_{r9}, (v) measurements of T_1 at the rising edge of v_r, and (vi) measurements of T_2 at the falling edge of v_r (adapted from [1]) (Figure 3.27).

3.3.3 Auxiliary Circuit-Based Method

Experiments are conducted on an SS-compensated WPT system configured with a semi-bridgeless rectifier. The system is powered by a 5 V DC input. The self-inductances of the transmitting and receiving coils are measured at 91.78 and 92.05 μH, respectively, with a coupling coefficient of 0.17 between the coils. The compensation capacitances for the system are set to 27.2 and 26.9 nF, respectively. Additionally, the ESRs of the transmitting and receiving resonators are 0.95 and 1.06 Ω, respectively. For the transmitter, a Hall-effect current sensor (LA12-10V21) is employed to measure the transmitter current. The output of the sensor is directly fed into the analog-to-digital conversion (ADC) pin of the transmitter controller, specifically the Texas Instruments' (TI's) F28379D Delfino Experimenter Kit. The phase delay introduced by the Hall-effect sensor, as well as the phase difference between the switching signal S_1 and the input fundamental voltage of the transmitter v_{t1} at critical frequencies, can be calibrated offline and compensated within the controller. This ensures that the equivalent input impedance Z_{in} remains purely resistive (i.e., v_{t1} and i_{t1} are in phase) at the critical frequencies (i.e., ω_A and ω_B). The proposed monitoring strategy is implemented in the transmitter controller. For the receiver, both switches of the semi-bridgeless rectifier are controlled and turned on by the receiver controller, specifically the TMS320F28335 Experimenter Kit. Since the phase differences between i_t and its fundamental component i_{t1} are negligible when the load is bypassed, the phase difference between v_{t1} and i_t can be effectively considered equivalent to the phase difference between v_{t1} and i_{t1}.

The switching frequency of the transmitter inverter is swept from its upper bound 115 kHz with a step size of −0.2 kHz. Figure 3.28a shows the waveforms of v_t, v_{t1}, i_t, and the output voltage of the DAC pin, which

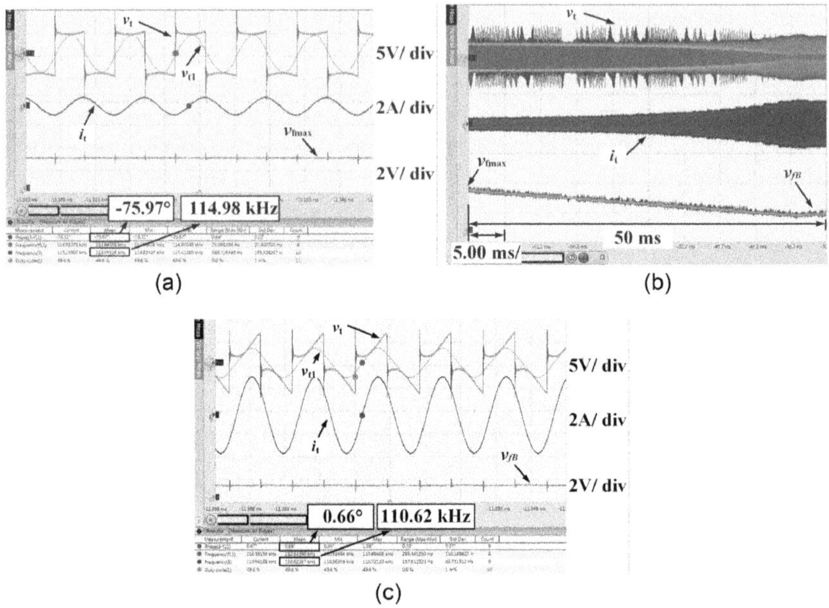

FIGURE 3.28
Waveforms of v_t, v_{t1}, i_t, and v_{ft} for the WPT system with $k=0.17$, operating from the upper bound frequency of 115 kHz to the critical frequency of 110.6 kHz: (a) steady-state performance at the upper bound frequency, (b) transient performance during the frequency sweep, and (c) steady-state performance at the critical frequency. (Adapted from reference [5].)

indicates the operating frequency (i.e., v_{fmax}) of the WPT system operating at the upper frequency bound. Here, i_t lags v_{t1} by approximately 75.97°. The reactance of the input impedance (i.e., $\mathbf{Im}(Z_{in})$) is inductive, which confirms the characteristic analysis presented in Table 3.2 and Figure 3.19. Subsequently, the operating frequency is swept from the upper bound to the critical frequency. The transmitter controller identifies the critical frequency at 110.6 kHz. The waveforms of v_t, v_{t1}, i_t, and v_{fB} during the frequency sweep and at steady-state operation at ω_B are shown in Figure 3.28b and c, respectively. Here, v_{fB} represents the output voltage of the DAC pin, corresponding to the operating frequency of the WPT system at the critical frequency above the resonant frequency. At steady state, v_{t1} and i_t are in phase. Using Equation (3.43), the coupling coefficient is estimated to be 0.1671, with a relative error of approximately 1.71%.

Next, the performance of sweeping the switching frequency from the lower bound of 90 kHz with a step size of 0.2 kHz is evaluated. The waveforms of v_t, v_{t1}, i_t, and v_{fmin} for the WPT system operating at the lower frequency bound are shown in Figure 3.29a. Here, i_t leads v_{t1} by approximately 86.72°. The reactance of the input impedance (i.e., $\mathbf{Im}(Z_{in})$) is capacitive, which confirms the characteristic analysis presented in Table 3.2 and Figure 3.19. The value of

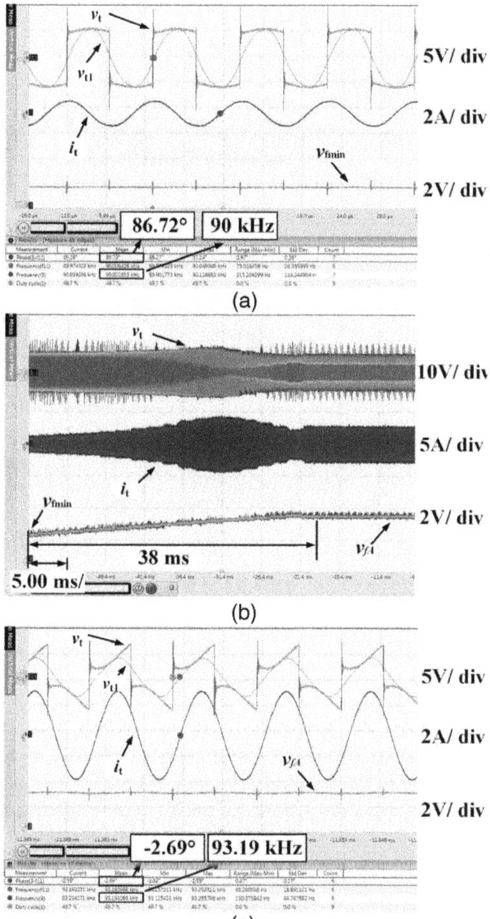

FIGURE 3.29
Waveforms of v_t, v_{t1}, i_t, and v_{ft} for the WPT system with $k = 0.17$, operating from the lower bound frequency of 90 kHz to the critical frequency of 93.2 kHz: (a) steady-state performance at the lower bound frequency, (b) transient performance during the frequency sweep, and (c) steady-state performance at the critical frequency. (Adapted from reference [5].)

v_{fmin} is 1.61 V, which indicates the operating frequency of the WPT system is at its lower bound of 90 kHz. Next, the operating frequency is swept from the lower bound to the critical frequency. The transmitter controller identifies the critical frequency below the resonant frequency as 93.2 kHz. The waveforms of v_t, v_{t1}, i_t, and v_{ft} during the search process are shown in Figure 3.29b. Here, v_{ft} is swept from 1.61 to 2.97 V in steps of 0.08 V and then adjusted from 2.97 to 2.89 V (i.e., v_{fA}) in the subsequent step. This indicates that the operating frequency of the WPT system changes from 90 to 93.4 kHz in steps of 0.2 kHz

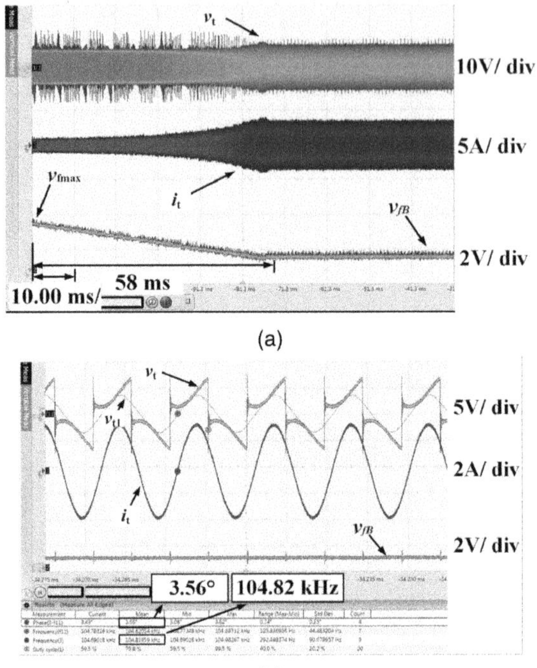

(a)

(b)

FIGURE 3.30
Waveforms of v_t, v_{t1}, i_t, and v_{ft} for the WPT system with $k = 0.075$, operating from the upper bound frequency of 110 kHz to the critical frequency of 104.8 kHz: (a) steady-state performance at the upper bound frequency and (b) transient performance during the frequency sweep. (Adapted from reference [5].)

and is then fine-tuned to the critical frequency of 93.2 kHz. The steady-state waveforms of v_t, v_{t1}, i_t, and v_{fp} for the WPT system operating at the critical frequency below the resonant frequency are shown in Figure 3.29c. Here, v_{t1} and i_t are in phase. Using Equation (3.42), the coupling coefficient is determined to be 0.1728, with a relative error of approximately 1.65%.

Then, the coupling coefficient of the coils is changed to be 0.076. The waveforms of v_t, v_{t1}, i_t, and v_{fp} of the WPT system during the frequency sweep and operating at the critical frequency are shown in Figure 3.30. At the steady state, v_{t1} and i_t are in phase. Using Equation (3.43), the coupling coefficient is calculated to be 0.0723, with a relative error of approximately 4.87%.

3.3.4 Discussion on the Empirical Results

From the experimental results, it is evident that both the heuristic algorithm and the auxiliary circuit-based monitoring strategies achieve high estimation accuracy. However, the auxiliary circuit-based non-resonant frequency strategy offers a distinct advantage: it can be implemented using cost-effective digital controllers, unlike heuristic algorithm-based methods, which require significant computational resources. The auxiliary circuit-based method ensures high estimation accuracy by bypassing the nonlinear rectifier load. Nevertheless, it is important to note that the monitoring parameters are significantly limited in the auxiliary circuit-based non-resonant frequency method. In this example, only the coupling coefficient or mutual inductance can be effectively monitored. In fact, the auxiliary circuit-based method is fundamentally a frequency sweeping approach, which, to some extent, can be viewed as a simplified version of the heuristic algorithm-based method. Its main advantage lies in its ability to mitigate the adverse effects caused by nonlinear rectifier loads. Fortunately, the derived estimation equations for the coupling coefficient are highly concise, and their hard implementation is remarkably straightforward. I observed that the identification inaccuracies in conventional monitoring strategies mainly stem from unmodeled parameters of the resonators and loads. To address this, I aimed to eliminate at least one influencing factor (either from the resonators or the loads) to improve accuracy. This led to the initial concept of the auxiliary circuit-based non-resonant frequency method. However, it became clear that this approach only partially addressed my vision. Recognizing the need to account for all potential uncertainties and disturbances, I concluded that a more robust solution was necessary. This motivated my exploration of AI techniques for monitoring key parameters in WPT systems.

3.4 Leveraging AI Technologies to Enhance Non-Resonant Frequency Methods

I categorize the applications of AI technologies for monitoring receiver-side parameters in typical WPT systems into two groups. The first category is purely data-driven, meaning the outputs of the AI-trained model are the precise parameters we intend to monitor. Since no analytical relationships between inputs and outputs are utilized, a substantial amount of training data is necessary to ensure accuracy. More importantly, the data must be collected across a wide range of operating conditions to achieve reliable results. The second category is semi-data-driven, meaning the outputs of the AI-trained model serve as compensation terms for the parameters we aim to monitor. The final monitored outputs are derived by combining the results from the analytical estimation model with the compensated outputs generated by the AI-trained model.

I acknowledge that the mechanisms underlying data-driven monitoring methods are not fully understood. However, previous studies [6,7] have demonstrated that AI-based monitoring strategies are independent of the sensitivities of the measured parameters. This implies that transmitter-side parameters can also be utilized to monitor receiver-side parameters with high accuracy. This raises an important question: If the measured parameters of a WPT system operating at resonant frequencies can be used to train the AI model, why is it necessary to use parameters measured at non-resonant frequencies for training? It is widely recognized that WPT systems are designed to operate at resonant frequencies to achieve high output power and efficiency. For this reason, I will defer the discussion of AI-based monitoring strategies to the next chapter, rather than addressing it here.

References

1. Y. Yang, "Precise modeling of nonlinear rectifier loads in wireless power transfer systems," *IEEE J. Emerg. Sel. Topics in Power Electron.*, vol. 11, no. 3, pp. 3574–3585, Jun. 2023.
2. J. Yin, D. Lin, T. Parisini, and S. Y. Hui, "Front-end monitoring of the mutual inductance and load resistance in a series–series compensated wireless power transfer system," *IEEE Trans. Power Electron.*, vol. 31, no. 10, pp. 7339–7352, Oct. 2016.
3. Y. Mao, Y. Yang, and K. Wang, "A primary-side fixed-frequency CC and CV output control for single-stage wireless battery chargers with series compensation receivers," *IEEE Trans. Transport. Electrific.*, vol. 10, no. 2, pp. 3407–3415, Jun. 2024.
4. Y. Yang, S. -C. Tan, and S. Y. R. Hui, "Front-end parameter monitoring method based on two-layer adaptive differential evolution for SS-compensated wireless power transfer systems," *IEEE Trans. Ind. Informat.*, vol. 15, no. 11, pp. 6101–6113, Nov. 2019.
5. Y. Yang, S. C. Tan, and S. Y. R. Hui, "Fast hardware approach to determining mutual coupling of series–series-compensated wireless power transfer systems with active rectifiers," *IEEE Trans. Power Electron.*, vol. 35, no. 10, pp. 11026–11038, Oct. 2020.
6. Y. Yang and H. Wu, "A robust primary-side hybrid data-driven load monitoring strategy for wireless power transfer systems," in *IEEE Energy Conversion Congress & Exposition (ECCE)*, Detroit, MI, USA, Oct. 2022, pp. 1–6.
7. Y. Yang, X. Zhang, and K. Wang, "A robust random forest regressor-based mutual inductance monitoring scheme for inductive power transfer systems," in *IEEE International Power Electronics and Application Conference and Exposition (PEAC)*, Guangzhou, China, Nov. 2022, pp. 171–176.

4

Resonant Frequency-Based Transmitter-Side Monitoring Strategies

4.1 Advantages of Resonant Frequency Monitoring over Non-Resonant Frequency Monitoring

The main advantage of resonant frequency monitoring methods over non-resonant frequency approaches is that the monitoring can be seamlessly integrated during charging, as most WPT systems operate at resonant frequencies in practice. WPT systems operate at resonant frequencies mainly due to the fixed frequency requirements necessary for compliance with popular wireless charging standards. However, the benefits of operating at resonant frequencies extend beyond compliance. A key advantage is the higher efficiency of the coupled resonators under typical operating conditions. Let me illustrate the efficiency difference between a typical WPT system operating at the resonant frequency and one operating at a non-resonant frequency with a simple example. The equivalent circuit of a typical SS-compensated WPT system in the fundamental frequency domain is depicted in Figure 4.1.

Based on the equivalent circuit in Figure 4.1,

$$v_{t1} = Z_t i_{t1} - j\omega M i_{r1} \tag{4.1}$$

$$j\omega M i_{t1} = \left(Z_r + R_{eq}\right) i_{r1} \tag{4.2}$$

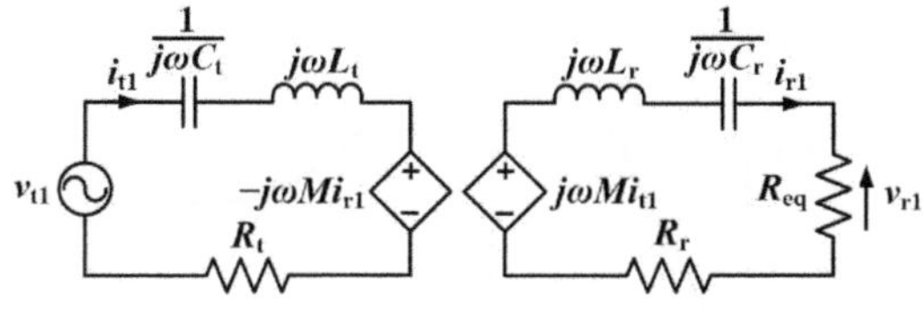

FIGURE 4.1
Equivalent circuit diagram in the fundamental frequency domain of a typical SS-compensated WPT system.

DOI: 10.1201/9781003468523-4

where $Z_t = j\omega L_t + \frac{1}{j\omega C_t} + R_t$ and $Z_r = j\omega L_r + \frac{1}{j\omega C_r} + R_r$. The transfer efficiency of the coupled resonators is

$$\eta = \frac{I_{r1}^2 R_{eq}}{I_{t1}^2 R_t + I_{r1}^2 \left(R_r + R_{eq}\right)} \tag{4.3}$$

By substituting Equation (4.2) into Equation (4.3) to eliminate the terms I_{t1} and I_{r1},

$$\eta = \frac{A_0}{\left(\frac{1}{\omega^2}\right)^2 + A_1\left(\frac{1}{\omega^2}\right) + A_2} \tag{4.4}$$

where $A_0 = \frac{R_{eq}}{R_t} M^2 C_r^2$, $A_1 = \left(R_r + R_{eq}\right)^2 C_r^2 - 2L_r C_r$, and $A_2 = L_r^2 C_r^2 + \left(\frac{R_r}{R_{eq}} + 1\right) A_0$. Based on Equation (4.4), the efficiency of the coupled resonators at various operating frequencies can be plotted, as illustrated in Figure 4.2. When the resonators operate at the maximum angular frequency as

$$\omega_{max} = \sqrt{\frac{-2}{A_1}} = \sqrt{\frac{2}{2L_r C_r - \left(R_{eq} + R_r\right)^2 C_r^2}} \tag{4.5}$$

the maximum efficiency of the resonators can be achieved as

$$\eta_{max} = \frac{4A_0}{4A_2 - A_1^2} = \frac{4R_{eq} M^2}{4\left(R_{eq} + R_r\right) M^2 - \left[\left(R_{eq} + R_r\right)^2 C_r - 2L_r\right]^2 R_t} \tag{4.6}$$

Apparently, the maximum angular frequency ω_{max} is higher than the resonant angular frequency ω_o

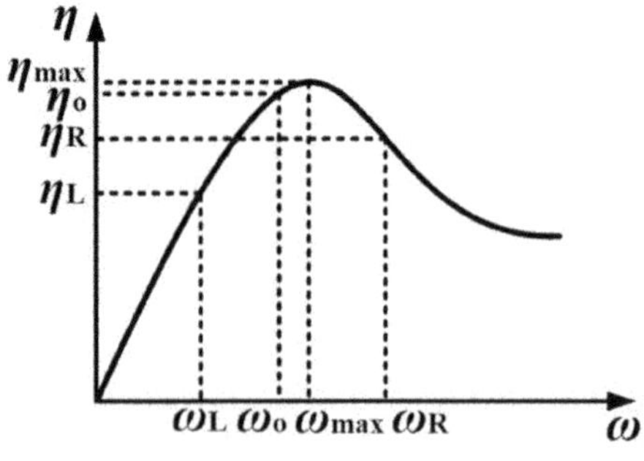

FIGURE 4.2
Efficiency curve of the SS-compensated resonators at various operating frequencies.

$$\omega_{\max} > \omega_o = \sqrt{\frac{1}{L_r C_r}} \tag{4.7}$$

The corresponding system efficiency at the resonant angular frequency ω_o is

$$\eta_o = \frac{R_{eq} M^2}{\left(R_{eq} + R_r\right) M^2 + R_t R_{eq}^2 L_r^2 C_r^2} \tag{4.8}$$

Compared to Equation (4.6), due to $4R_{eq}^2 L_r^2 C_r^2 + \left[\left(R_{eq} + R_r\right)^2 C_r - 2L_r\right]^2 > 0$,

$$\eta_o < \eta_{\max} \tag{4.9}$$

However, the difference between $\omega_{\max}$ and ω_o is generally negligible due to $\left(R_{eq} + R_r\right)^2 C_r^2 \approx 0$. Consequently, their efficiencies are also nearly identical.

Non-resonant frequency schemes typically require resonators to operate at frequencies biased by ±5% or more from their resonant frequency. Let me use frequency deviations of ±5% as an example (i.e., $\omega_L = 0.95\sqrt{\frac{1}{L_r C_r}}$ and $\omega_R = 1.05\sqrt{\frac{1}{L_r C_r}}$). Their corresponding efficiencies are

$$\eta_L = \frac{R_{eq} M^2}{\left(R_{eq} + R_r\right) M^2 + 1.1 R_t L_r C_r \left(R_{eq} + R_r\right)^2} \tag{4.10}$$

$$\eta_R = \frac{R_{eq} M^2}{\left(R_{eq} + R_r\right) M^2 + 0.91 R_t L_r C_r \left(R_{eq} + R_r\right)^2} \tag{4.11}$$

Obviously, η_R is greater than η_L. Additionally, operating the coupled resonators at frequencies above the resonant frequency is more favorable for achieving soft switching in the driving circuit. These explain why monitoring frequencies above the resonant frequency are typically preferred over those below the resonant frequency in non-resonant frequency methods. However, the efficiency η_R must be lower than the resonant efficiency η_o, as evidenced by Equations (4.8) and (4.11) and the following inequality condition:

$$0.95\left(1 + \frac{R_r}{R_{eq}}\right) > 0.95 \gg \frac{1}{\omega_o} \tag{4.12}$$

From the analysis above, it is evident that operating the WPT system at the resonant frequency offers advantages in terms of the efficiency of the coupled resonators compared to operating at non-resonant frequencies. However, it is important to note that the overall efficiency of a practical WPT system is also influenced by the efficiency of the power converters. By considering compatibility, efficiency, and monitoring sensitivity holistically, I believe a resonant frequency-based monitoring scheme is more advantageous than non-resonant frequency-based approaches.

4.2 Dynamic Response-Based Monitoring Scheme

4.2.1 Basic Analysis

To analyze the transient behavior, the Laplace-transformed equivalent circuit of a typical SS-compensated WPT system is illustrated in the phasor domain, as shown in Figure 4.3 [1–3].

Here, $V_t(s)$ and $V_r(s)$ represent the input and output voltages of the resonators, while $I_t(s)$ and $I_r(s)$ denote the transmitter and receiver currents, respectively. L_t and L_r are the self-inductances of the coils, and M is the mutual inductance. The transmitting-side and receiving-side compensation capacitors, C_t and C_r, are designed to satisfy the resonant condition $\omega_o = \frac{1}{\sqrt{L_t C_t}} = \frac{1}{\sqrt{L_r C_r}}$, where ω_o is the resonant angular frequency. The terms $j\omega M$, $j\omega L_t$, $j\omega L_r$, $\frac{1}{j\omega C_t}$, and $\frac{1}{j\omega C_r}$ represent the imaginary impedances of M, L_t, L_r, C_t, and C_r, respectively, resulting from the Laplace transformation, with ω being the operating angular frequency. Z_{eq} is the equivalent load impedance. For most load conditions, $Z_{eq} = \frac{8Z_o}{\pi^2}$ [4], where $Z_o = \frac{R_L}{sC_o R_L + 1}$, and R_L and C_o are the output resistance and capacitance of the RC load. To simplify the resonant tank, a capacitor in parallel with a resistor can be equivalently represented as an inductor in series with a resistor. Assuming the perturbation frequency is much lower than the operating angular frequency (i.e., $s \ll \omega$), the capacitors can be expressed as

$$\frac{1}{sC_t + j\omega C_t} = \frac{1 + js/\omega}{j\omega C_t\left(1 + s^2/\omega^2\right)} \approx \frac{1}{j\omega C_t} + \frac{s}{\omega^2 C_t} \tag{4.13}$$

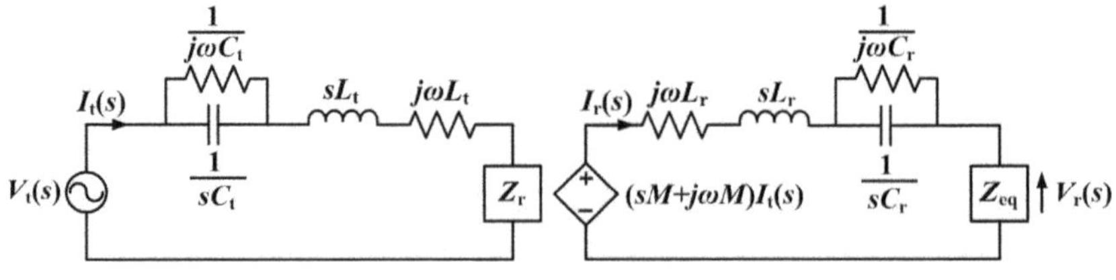

FIGURE 4.3
Laplace-transformed equivalent circuit of a typical SS-compensated WPT system in the phasor domain.

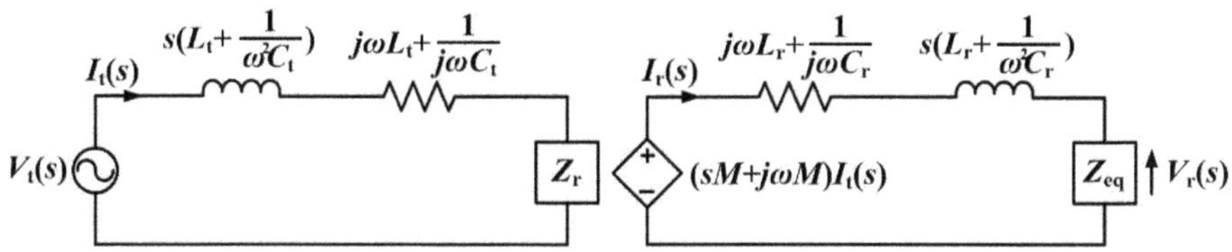

FIGURE 4.4
Simplified Laplace-transformed circuit of the SS-compensated WPT system in the phasor domain.

$$\frac{1}{sC_r + j\omega C_r} = \frac{1 + js/\omega}{j\omega C_r\left(1 + s^2/\omega^2\right)} \approx \frac{1}{j\omega C_r} + \frac{s}{\omega^2 C_r} \tag{4.14}$$

Consequently, the equivalent circuit can be simplified as shown in Figure 4.4. The reflected secondary-side impedance to the primary side is given by:

$$Z_r = \frac{\omega^2 M^2}{sL_r + \dfrac{s}{\omega^2 C_r} + j\omega L_r + \dfrac{1}{j\omega C_r} + Z_{eq}} \tag{4.15}$$

By applying Kirchhoff's voltage law, the system's equations operating at the resonant frequency (i.e., $\omega = \omega_o$) can be derived as follows:

$$\begin{cases} V_t(s) = \left(sL_t + \dfrac{s}{\omega_o^2 C_t} + Z_r\right) I_t(s) \\ (sM + j\omega_o M) I_t(s) = \left(sL_r + \dfrac{s}{\omega_o^2 C_r} + Z_{eq}\right) I_r(s) \end{cases} \tag{4.16}$$

Based on Equation (4.16), the transfer functions of the system in the transient state can be expressed as

$$\frac{V_r(s)}{V_t(s)} = \frac{\dfrac{8}{\pi^2}(\omega_o + s) M R_L}{4s^3 L_t L_r C_o R_L + 4s^2 L_t L_r + \dfrac{16}{\pi^2} s L_t R_L + (sC_o R_L + 1)(\omega_o + s)^2 M^2} \tag{4.17}$$

Since V_t and V_r are AC voltages, they are expressed in terms of V_{DC} and V_o, which represent the input and output DC voltages of the system, respectively, to provide a more intuitive interpretation of Equation (4.17). The relationships between $V_t(s)$, $V_r(s)$, $V_{DC}(s)$, and $V_o(s)$ are shown in Equation (4.18):

$$\begin{cases} V_t(s) = \dfrac{4}{\pi} V_{DC}(s) \\ V_r(s) = \dfrac{4}{\pi} V_o(s) \end{cases} \tag{4.18}$$

Substituting Equation (4.18) into Equation (4.17) yields

$$\frac{V_o(s)}{V_{DC}(s)}=\frac{\frac{8}{\pi^2}(\omega_o+s)MR_L}{4s^3L_tL_rC_oR_L+4s^2L_tL_r+\frac{16}{\pi^2}sL_tR_L+(sC_oR_L+1)(\omega_o+s)^2M^2} \tag{4.19}$$

To derive the small-signal transfer function, a small perturbation is added to the input DC voltage V_{DC}. As a result, Equation (4.19) can be rewritten as

$$\bar{V}_o+\hat{V}_o(s)=\frac{\frac{8}{\pi^2}(\omega_o+s)MR_L\left(\bar{V}_{DC}+\hat{V}_{DC}(s)\right)}{4s^3L_tL_rC_oR_L+4s^2L_tL_r+\frac{16}{\pi^2}sL_tR_L+(sC_oR_L+1)(\omega_o+s)^2M^2} \tag{4.20}$$

where $\bar{V}_o$ and $\bar{V}_{DC}$ are the average values, while $\hat{V}_o(s)$ and $\hat{V}_{DC}(s)$ are the perturbations of V_o and V_{DC}, respectively. Therefore, the transfer function of the small-signal model due to the perturbation of input DC voltage V_{DC} can be expressed based on Equations (4.19) and (4.20) as

$$\frac{\hat{V}_o(s)}{\hat{V}_{DC}(s)}=\frac{\frac{8}{\pi^2}(\omega_o+s)MR_L}{4s^3L_tL_rC_oR_L+4s^2L_tL_r+\frac{16}{\pi^2}sL_tR_L+(sC_oR_L+1)(\omega_o+s)^2M^2} \tag{4.21}$$

To derive the transfer function resulting from a small-signal perturbation $\hat{M}(s)$ on the mutual inductance, $V_o(s)=\bar{V}_o+\hat{V}_o(s)$ and $M=\bar{M}+\hat{M}(s)$ are substituted into Equation (4.19), where $\bar{M}$ represents the steady-state average value of M. By neglecting second-order small quantities, the transfer function of the small-signal model due to mutual inductance perturbation can be simplified using the Taylor expansion as follows:

$$\frac{\hat{V}_o(s)}{\hat{V}_{DC}(s)}=\frac{\frac{8}{\pi^2}(\omega_o+s)MR_L}{4s^3L_tL_rC_oR_L+4s^2L_tL_r+\frac{16}{\pi^2}sL_tR_L+(sC_oR_L+1)(\omega_o+s)^2M^2} \tag{4.22}$$

where $A=\frac{8V_{DC}R_L(\omega_o+s)}{\pi^2}$. Similarly, to derive the transfer function resulting from a small-signal perturbation $\hat{R}_L(s)$ on the load resistance R_L, $V_o(s)=\bar{V}_o+\hat{V}_o(s)$ and $R_o=\bar{R}_o+\hat{R}_o(s)$ are substituted into Equation (4.19), where $\bar{R}_o$ is the steady-state value of R_o. By applying the Taylor expansion and neglecting higher-order small terms, the transfer function of the small-signal model due to load resistance perturbation can be simplified as follows:

$$\frac{\hat{V}_o(s)}{\hat{R}_L(s)} = \frac{\frac{8}{\pi^2} V_{DC} M(\omega_o + s)\left[4s^2 L_t L_r + (\omega_o + s)^2 M^2\right]}{\left[4s^3 L_t L_r C_o \bar{R}_L + 4s^2 L_t L_r + \frac{16}{\pi^2} s L_t \bar{R}_L + \left(sC_o \bar{R}_L + 1\right)(\omega_o + s)^2 M^2\right]^2} \tag{4.23}$$

Based on Equation (4.16), the relationship between the transmitter-side current and the input DC voltage in the transient state can be expressed as

$$\frac{I_t(s)}{V_{DC}(s)} = \frac{\frac{4}{\pi}\left(2s^2 L_r C_o R_L + 2sL_r + \frac{8}{\pi^2} R_L\right)}{4s^3 L_t L_r C_o R_L + 4s^2 L_t L_r + \frac{16}{\pi^2} s L_t R_L + (sC_o R_L + 1)(\omega_o + s)^2 M^2} \tag{4.24}$$

where $I_t(s)$ represents the peak value of the transmitter current. Similarly, the small-signal models describing the effects of variations in V_{DC}, M, and R_L on I_t can be derived from Equation (4.24) as follows:

$$\frac{\hat{I}_t(s)}{\hat{V}_{DC}(s)} = \frac{\frac{4}{\pi}\left(2s^2 L_r C_o R_L + 2sL_r + \frac{8}{\pi^2} R_L\right)}{4s^3 L_t L_r C_o R_L + 4s^2 L_t L_r + \frac{16}{\pi^2} s L_t R_L + (sC_o R_L + 1)(\omega_o + s)^2 M^2} \tag{4.25}$$

$$\frac{\hat{I}_t(s)}{\hat{M}(s)} = \frac{-\frac{8}{\pi} \bar{M} V_{DC}\left(2s^2 L_r C_o R_L + 2sL_r + \frac{8}{\pi^2} R_L\right)(sC_o R_L + 1)(\omega_o + s)^2}{\left[4s^3 L_t L_r C_o R_L + 4s^2 L_t L_r + \frac{16}{\pi^2} s L_t R_L + (sC_o R_L + 1)(\omega_o + s)^2 \bar{M}^2\right]^2} \tag{4.26}$$

$$\frac{\hat{I}_t(s)}{\hat{R}_L(s)} = \frac{\frac{32}{\pi^3} V_{DC} (\omega_o + s)^2 M^2}{\left[4s^3 L_t L_r C_o \bar{R}_L + 4s^2 L_t L_r + \frac{16}{\pi^2} s L_t \bar{R}_L + \left(sC_o \bar{R}_L + 1\right)(\omega_o + s)^2 M^2\right]^2} \tag{4.27}$$

where $\hat{I}_t(s)$ is the perturbation of I_t.

To analyze the dynamic stability of the system under perturbations, consider the small-signal model that describes how variations in V_{DC} affect I_t. According to Equation (4.25), the coefficients of s^3 and s^2 in the denominator of the transfer function are negligible compared to those of s^1 and s^0. Therefore, the overshoot $I_{tsettle}$ and settling time T_{settle} of the transient waveforms of i_t in the time domain can be derived as

$$I_{tsettle} = \frac{32 R_L \hat{V}_{DC}}{\pi^3 \omega_o^2 M^2} \tag{4.28}$$

$$T_{settle} = \frac{4\left(\frac{16}{\pi^2} L_t R_L + \omega_o^2 M^2 C_o R_L + 2\omega_o M^2\right)}{\omega_o^2 M^2} \tag{4.29}$$

where $\hat{V}_{DC}$ represents the amplitude of $\hat{V}_{DC}(s)$. Based on Equations (4.28) and (4.29), the real-time values of M and R_L in the WPT system can be estimated at the resonant frequency using only transmitter-side measurements as

$$\tilde{M} = 2\sqrt{2}\sqrt{\frac{\omega_o T_{settle}\hat{V}_{DC} - 8\hat{V}_{DC} - 2\pi\omega_o L_t I_{tsettle}}{\pi^3\omega_o^3 C_o I_{tsettle}}} \tag{4.30}$$

$$\tilde{R}_L = \frac{\omega_o T_{settle}\hat{V}_{DC} - 8\hat{V}_{DC} - 2\pi\omega_o L_t I_{tsettle}}{4\omega_o C_o \hat{V}_{DC}} \tag{4.31}$$

Similarly, the real-time values of M and R_L can be estimated at the resonant frequency using the model in Equation (4.27) based on the perturbation $\hat{R}_L(s)$ as

$$\tilde{M} = 4\sqrt{2}\sqrt{\frac{V_{DC}\hat{R}_L}{\pi^3\omega_o^2 I_{tsettle}}} \tag{4.32}$$

$$\tilde{R}_L = \frac{\omega_o T_{settle}V_{DC}\hat{R}_L - 8V_{DC}\hat{R}_L}{4\omega_o V_{DC}C_o\hat{R}_L + 2\pi\omega_o L_t I_{tsettle}} + \hat{R}_L \tag{4.33}$$

where $\hat{R}_L$ represents the amplitude of $\hat{R}_L(s)$. Notably, the value of Equation (4.33) corresponds to the evaluation of R_L in steady state after the dynamic process, whereas the initial value of R_L before the perturbation can be determined by subtracting $\hat{R}_L$ from Equation (4.33). Similarly, real-time values of M and R_L can be estimated at the resonant frequency using the model in Equation (4.26) based on the perturbation $\hat{M}(s)$ by solving the following two equations:

$$I_{tsettle} = \frac{64R_L V_{DC}\hat{M}}{\pi^3\omega_o^2\bar{M}^3} \tag{4.34}$$

$$T_{settle} = \frac{4\left(\frac{16}{\pi^2}L_t R_L + \omega_o^2\bar{M}^2 C_o R_L + 2\omega_o\bar{M}^2\right)}{\omega_o^2\bar{M}^2} \tag{4.35}$$

where $\hat{M}$ is the amplitude of $\hat{M}(s)$. The $\tilde{M}$ obtained from the solution represents the steady-state value before the perturbation. The steady-state M after the perturbation needs to be adjusted by adding $\hat{M}$.

It is important to note that, in practice, the captured data is discrete, which may cause the measured overshoot $I_{tsettle}$ and settling time T_{settle} to deviate from their actual values [5]. To minimize deviations that could compromise estimation accuracy, values below zero and those that deviate significantly from the preceding and subsequent five peaks are filtered out. This process establishes a robust baseline for identifying peak values. Subsequently, the peaks corresponding to the steady-state condition are determined. In [5], a sequence of 20 consecutive peaks with an average deviation of less than 1% is considered to represent the steady state. The overshoot $I_{tsettle}$ is defined as the difference between the average value of the first steady-state period and the average value of the last steady-state period. To determine the settling time T_{settle}, the onset of dynamic changes is identified by locating the first peak that deviates by 20% from its preceding peak. Following the definition of T_{settle}, the target peak is identified as the one with a deviation of less than 2% from the final steady-state value. The time interval between the starting point of dynamic changes and the target peak is designated as T_{settle}. To ensure accurate identification of peaks during the dynamic process, the deviation between the target peak and the subsequent peak is constrained to within 5%.

4.2.2 Experimental Verification

Experiments are performed on the WPT setup, powered by a KIKUSUI PWR2001H DC power supply. The transmitter-side full-bridge inverter employs an IMW120R020M1HXKSA1 SiC FET. Switching signals are generated using a digital signal processor (DSP) TMS320F28335. A diode bridge rectifier is implemented with four APT30S20BG diodes.

For the WPT system with a mutual inductance (i.e., M) of 27.145 μH and a load resistance (i.e., R_L) of 36.05 Ω, the input DC voltage (i.e., V_{DC}) is stepped up from 0 to 10 V. Figure 4.5 demonstrates the process of determining the overshoot $I_{tsettle}$ and settling time T_{settle}. In Figure 4.5, the circles denote the peaks of the transmitting current I_t, corresponding to the upper envelope of the transmitting current waveform i_t. The triangles indicate the first steady-state period, while the squares represent the last steady-state period. The starting point is marked by a multiplication sign, and the target peak is marked by a plus sign. The overshoot $I_{tsettle}$ is defined as the difference between the average value of the last steady-state period and the average value of the first

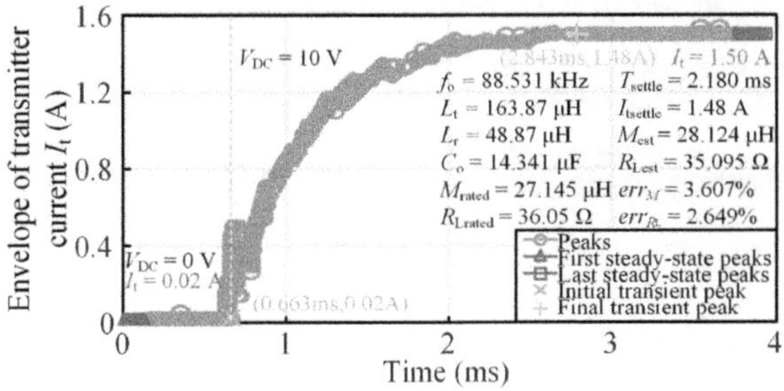

FIGURE 4.5
Identification process of the overshoot $I_{tsettle}$ and the settling time T_{settle} in practice. (Adapted from reference [5].)

steady-state period. The settling time T_{settle} is calculated as the time interval between the starting point and the target peak.

The experimental waveforms of v_t and i_t during the startup period for different values of M and R_L are shown in Figure 4.6. The parameters for each experimental group, including the switching frequency f_o, the inductances of the transmitter and receiver coils (i.e., L_t and L_r), the filtering capacitor C_o, the mutual inductance M, and the load resistance R_L, are labeled in the figures. In the experiments, the soft-starting process is simulated by transitioning the full-bridge inverter from a state where all four switches are turned off to a half-bridge operation mode and finally to a full-bridge operation mode. This process is equivalent to stepping the input voltage V_{DC} from 0 to 5 V, followed by a second step to 10 V. Cases 1 and 2 operate under weak coupling conditions, with a coupling coefficient of k=0.09, mutual inductance M=8.435 μH, and load resistances R_L=4.7 Ω and R_L=6.863 Ω, respectively. In contrast, Cases 3 and 4 operate under strong coupling conditions, with a coupling coefficient of k=0.28 and the same respective load resistances R_L=4.7 Ω and R_L=6.863 Ω.

As shown in Figure 4.6a, V_{DC} remains at 0 V until 0.0924 ms, after which it steps up to 5 V, initiating the system startup. This step-up results in a settling time of T_{settle}=0.5242 ms and an overshoot of $I_{tsettle}$=1.2240 A. Using Equations (4.30) and (4.31), the corresponding mutual inductance and load resistance are estimated as M=7.940 μH and R_L=4.528 Ω. At 1.1101 ms, V_{DC} steps up again to 10 V, leading to a new T_{settle} of 0.4795 ms

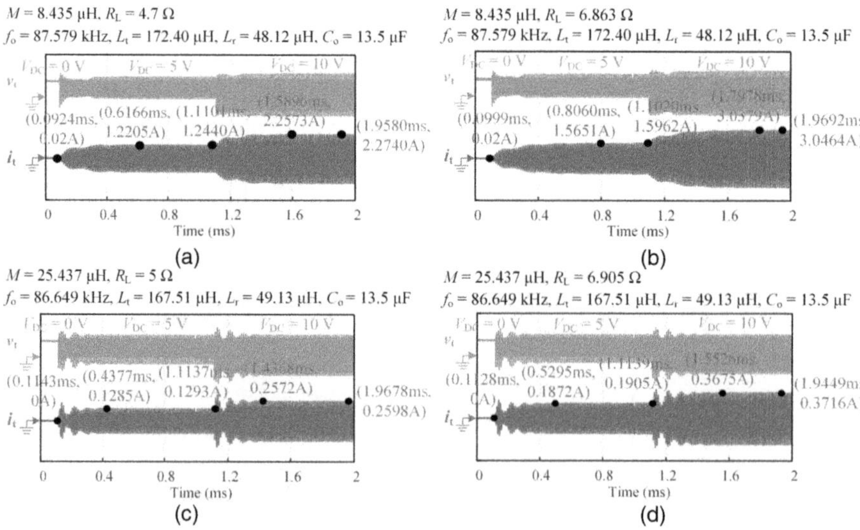

FIGURE 4.6
Experimental waveforms of the transmitter input voltage v_t and current i_t in response to variations in V_{DC}, for different values of mutual inductance M and load resistance R_L: (a) M=8.435 μH and R_L=4.7 Ω, (b) M=8.435 μH and R_L=6.863 Ω, (c) M=25.437 μH and R_L=5 Ω, and (d) M=25.437 μH and R_L=6.905 Ω. (Adapted from reference [5].)

and an $I_{tsettle}$ of 1.0300 A, yielding estimated values of M=8.608 μH and R_L=4.478 Ω. Throughout the startup process, the average estimated values are M=8.274 μH and R_L=4.503 Ω, corresponding to errors of –1.909% and –4.191%, respectively, when compared to the actual values of 8.435 μH and 4.7 Ω. As depicted in Figure 4.6b, V_{DC} remains at 0 V until 0.0999 ms, after which it steps up to 5 V. This initial step-up results in a settling time T_{settle} of 0.7061 ms and an overshoot $I_{tsettle}$ of 1.5762 A. Based on these measurements and using Equations (4.30) and (4.31), the estimated mutual inductance is calculated to be 8.372 μH, and the load resistance is estimated at 6.483 Ω. At 1.1020 ms, V_{DC} steps up again to 10 V, yielding a new T_{settle} of 0.6958 ms and an $I_{tsettle}$ of 1.4502 A. These values correspond to an estimated M of 8.938 μH and R_L of 6.798 Ω. Over the entire startup process, the average estimated values are M=8.655 μH and $R_{L=}$6.6405 Ω. Compared to the practical values of 8.435 μH and 6.863 Ω, these estimates reflect errors of 2.608% and –3.242%, respectively. As exhibited in Figure 4.6c, V_{DC} remains at 0 V until 0.1143 ms, after which it steps up to 5 V, initiating the system startup. This step-up results in a settling time of T_{settle}=0.3234 ms and an overshoot of $I_{tsettle}$=0.1293 A. Using Equations (4.30) and (4.31), the corresponding mutual inductance and load resistance are estimated as M=26.493 μH and R_L=5.213 Ω. At 1.1137 ms, V_{DC} steps up again to 10 V, leading to a new T_{settle} of 0.3231 ms and an $I_{tsettle}$ of 0.1305 A, yielding estimated values of M=26.345 μH and R_L=5.203 Ω. Throughout the startup process, the average estimated values are M=26.419 μH and R_L=5.208 Ω, corresponding to errors of 3.861% and 4.160%, respectively, when compared to the actual values of 25.437 μH and 5 Ω. As illustrated in Figure 4.6d, V_{DC} remains at 0 V until 0.1128 ms, after which it steps up to 5 V. This initial step-up results in a T_{settle} of 0.4167 ms and an $I_{tsettle}$ of 0.1905 A. Based on these measurements and using Equations (4.30) and (4.31), the estimated M is calculated to be 24.748 μH, and R_L is estimated at 6.702 Ω. At 1.1139 ms, V_{DC} steps up again to 10 V, yielding a new T_{settle} of 0.4387 ms and an $I_{tsettle}$ of 0.1811 A. These values correspond to an estimated M of 26.210 μH and R_L of 7.146 Ω. Over the entire startup process, the average estimated values are M=25.479 μH and R_L=6.924 Ω. Compared to the practical values of 25.437 μH and 6.905 Ω, these estimates reflect errors of 0.165% and 0.275%, respectively.

The relative errors for the four cases are exhibited in Figure 4.7, with the average relative errors of M and R_L remaining below 2.136% and 2.967%, respectively, across all cases. As shown in Figure 4.7a, the relative error in the estimated mutual inductance M decreases as M increases, given a constant load resistance R_L. Similarly, Figure 4.7b demonstrates that the relative error in the estimated load resistance R_L decreases as R_L increases when mutual inductance M remains unchanged. Moreover, the relatively low average relative error suggests that multiple measurements help mitigate inaccuracies. This indicates that conducting multiple estimations during the soft-starting process can further enhance accuracy and reduce estimation errors.

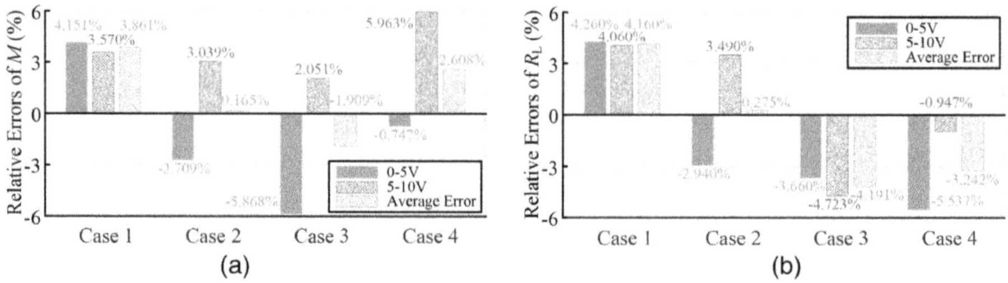

FIGURE 4.7
The relative errors for the four cases of (a) M and (b) R_L. (Adapted from reference [5].)

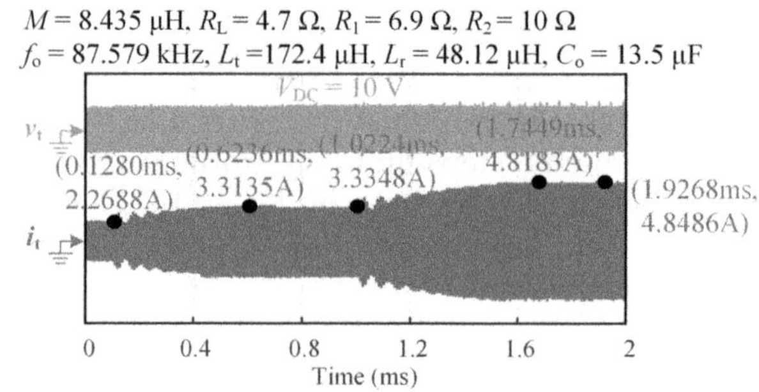

FIGURE 4.8
Experimental waveforms of v_t and i_t under varying load conditions. (Adapted from reference [5].)

The experimental waveforms of the transmitter input voltage v_t and current i_t in response to load resistance perturbations are shown in Figure 4.8. The experimental parameters are labeled on the figure, including the switching frequency f_o, the inductances of the transmitter and receiver coils (i.e., L_t and L_s), the filtering capacitor C_o, the mutual inductance M, the load resistance R_L, and the input DC voltage V_{DC}. As illustrated in Figure 4.8, V_{DC} remains constant at 10 V, and M remains fixed at 8.435 μH. Initially, R_L is set to 4.7 Ω until 0.1280 ms, after which it steps up to 6.9 Ω. This step-up results in a settling time T_{settle} of 0.4956 ms and an overshoot $I_{tsettle}$ of 1.0660 A. Based on these measurements and using Equations (4.32) and (4.33), the estimated mutual inductance is calculated to be 8.738 μH, and the load resistance is estimated at 7.088 Ω. At 1.0224 ms, R_L steps up again to 10 Ω, yielding a new settling time of 0.7225 ms and an overshoot of 1.5138 A. These values correspond to an estimated M of 8.716 μH and R_L of 10.274 Ω. Across the two dynamic processes involving perturbations in R_L, the average estimation errors for M and R_L are 3.462% and 2.732%, respectively. These results highlight the accuracy and reliability of the proposed method in identifying mutual inductance and load resistance under varying operating conditions.

In [5], the dynamic response-based scheme is evaluated against the conventional non-resonant frequency scheme in terms of steady-state monitoring accuracy across the eight cases shown in Figure 4.6. The dynamic response-based scheme achieves an average relative error of 2.967%, significantly lower than the 9.867% average relative error of the conventional non-resonant frequency scheme.

4.2.3 Discussion on the Dynamic Response-Based Scheme

The main advantage of the dynamic response-based monitoring scheme lies in its ability to monitor receiver-side parameters in real time during operation, making it especially suitable for dynamic WPT systems. However, this approach also encounters significant challenges, such as ensuring accurate measurements under rapidly changing conditions and minimizing the effects of noise and disturbances on the accuracy of overshoot and settling time estimations. Another practical issue that cannot be overlooked is that changes in input DC voltage, mutual inductance, and load conditions may not occur as instantaneous step changes in real-world scenarios. For example, the input DC voltage often requires a rectifier to convert AC to DC, along with a controller to regulate the rectifier. The feedback control loop, together with the filter capacitor in the rectifier, can cause the input DC voltage to change gradually rather than abruptly. This leads to significantly more complex estimation equations. Similar challenges arise with variations in mutual inductance and load conditions. Therefore, I recommend developing more comprehensive estimation equations that account for WPT systems with diverse compensation topologies and dynamic response characteristics.

4.3 Harmonic-Based Monitoring Scheme

One of the major challenges of conventional non-resonant frequency monitoring schemes is their high sensitivity near the resonant frequency. To address this issue, a harmonic-based monitoring scheme is proposed [6]. The key advantage of the harmonic-based monitoring scheme over conventional non-resonant frequency monitoring methods is that it enables real-time monitoring during the WPT system operation, as harmonic frequencies are positioned farther from the resonant frequency. The primary challenge of the harmonic-based monitoring scheme is the need for current sensors with a wide bandwidth to accurately measure the harmonic components.

4.3.1 Basic Analysis

In practical WPT systems operating across a wide range of load and mutual inductance conditions, significant harmonic components are present, especially when the transmitter input voltage is a square waveform generated by a full-bridge inverter. The information embedded in the harmonic components, such as third-order, fifth-order, and even higher-order harmonics, can be utilized to formulate sufficient estimation equations for transmitter-side monitoring of critical parameters on the receiver side. To illustrate the harmonic-based monitoring scheme, I will use an SS-compensated WPT system as an example.

The equivalent input impedance of the coupled SS-compensated resonators is

$$Z_{\text{in}n} = R_{\text{t}n} + jX_{\text{t}n} + \frac{n^2\omega^2 M_n^{\,2}}{R_{\text{eq}} + R_{\text{r}n} + jX_{\text{r}n}} \tag{4.36}$$

where n represents the harmonic order. $R_{\text{t}n}$ and $R_{\text{r}n}$ represent the ESRs of the transmitter and receiver resonators at the operating angular frequency $n\omega$, respectively. M_n denotes the mutual inductance at $n\omega$. R_{eq} is the equivalent nonlinear load resistance [4]. $X_{\text{t}n}$ and $X_{\text{r}n}$ represent the imaginary components of the equivalent impedances of the transmitter and receiver resonators, respectively, and can be expressed as

$$X_{\text{t}n} = n\omega L_{\text{t}n} - \frac{1}{n\omega C_{\text{t}n}} \tag{4.37}$$

$$X_{\text{r}n} = n\omega L_{\text{r}n} - \frac{1}{n\omega C_{\text{r}n}} \tag{4.38}$$

where $L_{\text{t}n}$ and $L_{\text{r}n}$ denote the self-inductances of the transmitter and receiver coils at the operating angular frequency $n\omega$, respectively. Similarly, $C_{\text{t}n}$ and $C_{\text{r}n}$ represent the compensated capacitors in the transmitter and receiver resonators at $n\omega$, respectively. The real and imaginary components of $Z_{\text{in}n}$ can be expressed as

$$R_{\text{in}n} = R_{\text{t}n} + \frac{n^2\omega^2 M_n^{\,2}\left(R_{\text{eq}} + R_{\text{r}n}\right)}{\left(R_{\text{eq}} + R_{\text{r}n}\right)^2 + X_{\text{r}n}^{\,2}} \tag{4.39}$$

$$X_{\text{in}n} = X_{\text{t}n} - \frac{n^2\omega^2 M_n^{\,2} X_{\text{r}n}}{\left(R_{\text{eq}} + R_{\text{r}n}\right)^2 + X_{\text{r}n}^{\,2}} \tag{4.40}$$

Based on Equations (4.39) and (4.40), R_{eq} can be derived as

$$R_{\text{eq}} = \frac{\left(R_{\text{in}n} - R_{\text{t}n}\right)X_{\text{r}n}}{X_{\text{t}n} - X_{\text{in}n}} - R_{\text{r}n} \tag{4.41}$$

By substituting Equation (4.41) into Equation (4.39) or Equation (4.40), when $n\omega \neq \omega_{\text{or}}$, where ω_{or} is the resonant angular frequency of the receiving resonator, the mutual inductance can be determined by

$$M_n = \sqrt{\frac{\left[\left(\frac{R_{inn} - R_{tn}}{X_{tn} - X_{inn}}\right)^2 + 1\right] X_{rn}\left(X_{tn} - X_{inn}\right)}{n^2\omega^2}} \tag{4.42}$$

Therefore, if the harmonic components can be accurately extracted, the mutual inductance of the WPT system can still be estimated even when the system operates at the resonant frequency. This enables real-time estimation of mutual inductance without shifting the switching frequency from the resonant frequency. By utilizing a lookup table that maps the relationship between M_n and M_1, the fundamental mutual inductance M_1 can be determined.

4.3.2 Experimental Validation

Experiments are conducted on an SS-compensated WPT system with the main parameters given in Table 4.1. The parameter values are measured using a Bode 100 network analyzer. A combination of the Aigtek ATA-1220D wide-band amplifier and the Tektronix AFG 31000 series arbitrary function generator is used to generate sinusoidal signals at the desired frequencies.

The estimated coupling coefficients of the WPT system based on Equation (4.42) are plotted in Figure 4.9. The corresponding relative estimation errors are shown in Figure 4.10. The estimations demonstrate high accuracy when $k > 0.2$. However, under weaker coupling conditions, the estimation errors become significant. This is primarily because the harmonic components are relatively small when the coils are weakly coupled, making them more susceptible to noise in the harmonic extraction process. For example, a 0.5% measurement error from the sensor can result in a 5% estimation error when $k = 0.154$. Fortunately, the coupling coefficients in most "Qi" or "Qi2"-compatible WPT systems typically range between 0.2 and 0.6, where the harmonic-based monitoring scheme achieves high estimation accuracy, particularly at the resonant frequency.

TABLE 4.1

Key Parameters of the WPT System Used to Validate the Harmonic-Based Scheme

Parameter	Value	Parameter	Value
Transmitter inductance L_t	78.19 μH	Receiver inductance L_r	77.31 μH
Transmitter capacitance C_t	29.72 nF	Receiver capacitance C_r	29.72 nF
Transmitter resistance R_t	0.5 Ω	Receiver resistance R_r	0.2 Ω
Resonant frequency of the transmitter resonator f_{ot}	104.4 kHz	Resonant frequency of the receiver resonator f_{or}	105 kHz
Input DC voltage V_{DC}	5 V	Output DC capacitance C_o	10 μF
Load resistance R_L	2 Ω		

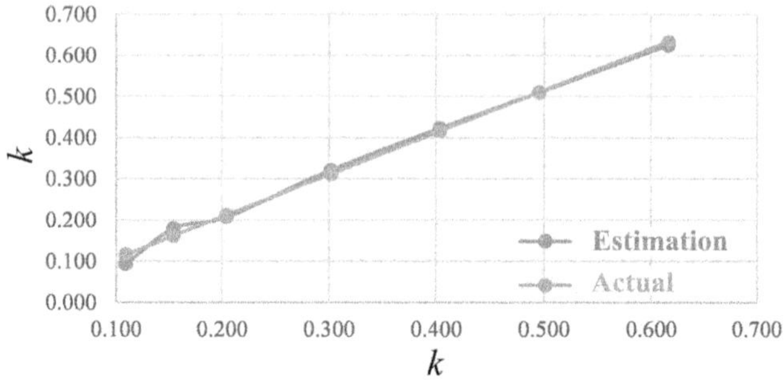

FIGURE 4.9
Comparison of the actual coupling coefficient and the estimated coupling coefficient derived from the harmonic-based monitoring scheme. (Adapted from reference [6].)

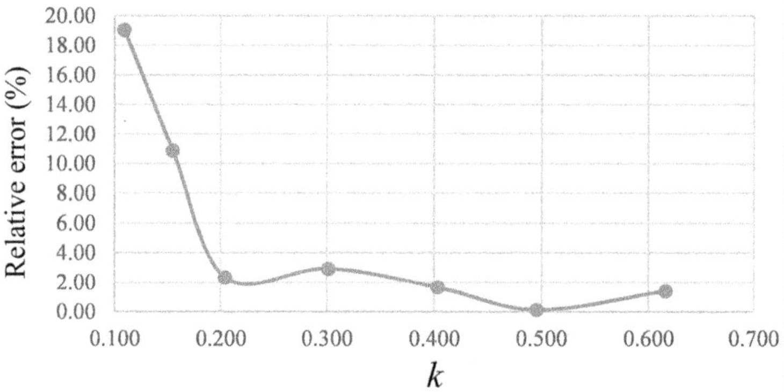

FIGURE 4.10
Relative estimation errors of the coupling coefficient using the harmonic-based monitoring scheme. (Adapted from reference [6].)

4.3.3 Discussion on the Harmonic-Based Scheme

In my view, the core concept of the harmonic-based monitoring scheme closely resembles that of non-resonant frequency monitoring methods. The accuracy of this approach can be quite low when the harmonic components are minimal, but it achieves high accuracy when these components are more pronounced. The extent of harmonic components in a WPT system is influenced by multiple factors, including mutual coupling and load conditions. A promising application of this method could be in fault detection, foreign object identification, or aging diagnostics. For example, an aged battery load may increase the equivalent load resistance, leading to a higher level of harmonic components in certain WPT systems with specific compensation networks.

4.4 AI-Based Monitoring Scheme

AI-based monitoring schemes can be either purely data-driven or a hybrid approach that combines both model-driven and data-driven methods. In simple terms, purely data-driven schemes estimate the receiving-side parameters directly without relying on prior knowledge of the WPT system. These approaches output mutual inductances or load conditions directly rather than their compensated values. On the contrary, hybrid schemes estimate the compensated values, which are then combined with the monitoring parameters derived from theoretical models.

4.4.1 Purely Data-Driven Scheme

For purely data-driven strategies, a substantial amount of data is required due to the nonlinear relationships between transmitter-side measurements, mutual inductances, and load conditions, especially when accounting for parameter drifts or external disturbances. Therefore, the challenge of managing large data volumes must be resolved to enable effective monitoring strategies in local controllers. It is widely recognized that acquiring a substantial amount of data from practical measurements is often difficult in real-world scenarios. To tackle this issue, a hybrid data approach has been proposed in [7]. Another critical consideration is the selection of appropriate AI algorithms to train and validate data-driven models using limited datasets (even when employing hybrid data), as local controller units have constrained data storage capacity. So far, we have explored the use of Multilayer Perceptron (MLP), Random Forest (RF), and Extreme Gradient Boosting (XGBoost) to train and validate purely data-driven models for transmitter-side monitoring and control WPT systems [7–10].

4.4.1.1 Overview of the MLP-Based Method

In [7], the hybrid data we employ consists of both simulation and experimental data. The simulation data can be quickly generated in large quantities, while the experimental data captures real-world disturbances. This combined dataset is used to train an MLP with one input layer, six hidden layers, and one output layer. The main advantage of using simulation data is that it can be generated rapidly and in large quantities at a relatively low cost. Additionally, processing simulation data is generally simpler compared to experimental data. However, a significant limitation of simulation data is that it may not fully capture certain unmodeled disturbances, such as temperature variations or current deviations. Conversely, experimental data can account for these disturbances, but acquiring large volumes of experimental data for training the AI models is often costly and time-consuming.

Consequently, combining the strengths of both data types through a hybrid approach offers a promising solution to leverage their respective benefits.

An MLP is a network composed of neurons arranged in layers, where computations are carried out sequentially through these layers. It is a classic feed-forward neural network with multiple hidden layers, meaning information flows in one direction without looping back [11]. Neurons in adjacent layers are fully connected, while neurons within the same layer are not linked. Each neuron performs a mathematical operation, applying an activation function to the weighted sum of its inputs plus a bias term. MLPs are effective nonlinear function approximators that use gradient descent to minimize errors in function approximation. The core principles of backpropagation and gradient descent have remained largely unchanged since the 1980s. A significant improvement in recent years is the shift from sigmoid to Rectified Linear Unit (ReLU) activation functions [12]. For small datasets, using ReLU can be more impactful than fine-tuning hidden layer weights [13]. MLPs perform well in classification and regression tasks when supported by appropriate resources and engineering practices. Their gradient-based learning approach is effective across various machine learning (ML) applications [14]. Compared to other supervised ML models like Convolutional Neural Networks (CNNs) and Recurrent Neural Networks (RNNs), MLPs offer the advantages of (i) accuracy and robustness, (ii) versatility and flexibility, (iii) scalability and adaptability, (iv) access to internal data, and (v) suitability for structured datasets [15].

Figure 4.11 exhibits the training and validation workflow of the MLP algorithm for the hybrid data-driven load monitoring strategy. Initially, the fundamental components V_{t1} and I_{t1}, and the mutual inductance M, are extracted from both simulation and experimental data. The data weighting factor (i.e., λ) determines the proportion between the simulation and experimental data. For example, if the total number of data points is N, the simulation data comprises λN, and the experimental data comprises $(1-\lambda)N$. The raw measured parameters cannot be used directly, so data preprocessing steps such as removing invalid data and normalization are performed to create a clean dataset, which is then split into (i) a training set and (ii) a test set. The training set is used to train the MLP with an optimized structure and parameters. During training, the training set is further divided into $(i-1)$ a training subset and $(i-2)$ a validation subset. Both subsets are used to evaluate the MLP. However, the training subset is also used for parameter selection, with updates made through learning feedback. The MLP performance is evaluated using both subsets to compute error metrics. These metrics are primarily used to check for overfitting and convergence. If the model has not converged, the MLP parameters must be adjusted. If convergence is achieved, the MLP model is exported for evaluation using the test set. Both the quality and adaptability of the evaluated MLP model are then checked for further optimization of its structure and parameters.

The resulting MLP model serves as an advanced load monitor, offering greater robustness against parameter drifts and measurement errors

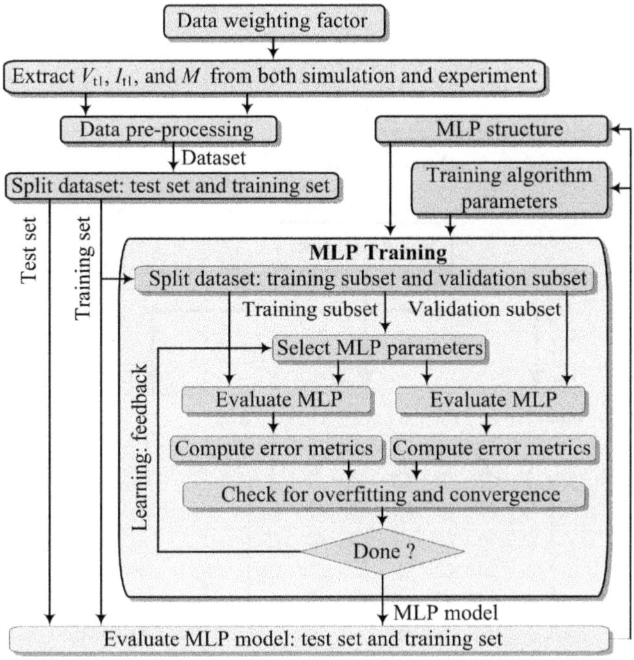

FIGURE 4.11
Training and validation workflow of the MLP based on hybrid data for load monitoring. (Adapted from reference [7].)

compared to conventional model-based load monitoring schemes. The schematic diagram of the MLP-based load monitoring strategy is depicted, as shown in Figure 4.12. The measured parameters from the practical setup also require preprocessing. The normalized values of V_{t1}, I_{t1}, and M are fed as inputs into the trained MLP model, with the estimated R_L as the output.

The MLP is designed to approximate any continuous function and can address problems that are not linearly separable. The input layer of the MLP receives the parameters of the WPT system for further processing, while the output layer generates the estimated R_L. The hidden layers, positioned between the input and output layers, serve as the core computational engine of the MLP. Data flows from the input layer to the output layer through a feedforward network. The neurons in the MLP are trained using the Adam optimizer. The MLP is configured with six hidden layers, each containing 128 neurons with a ReLU activation function.

4.4.1.2 Experimental Validation of the MLP-Based Method

Experiments are conducted on the SS-compensated WPT system with key parameters given in Table 4.2. The MLP consists of 128 neurons per layer, and the Adam optimizer is employed with a learning rate of 10^{-4}. For training, five

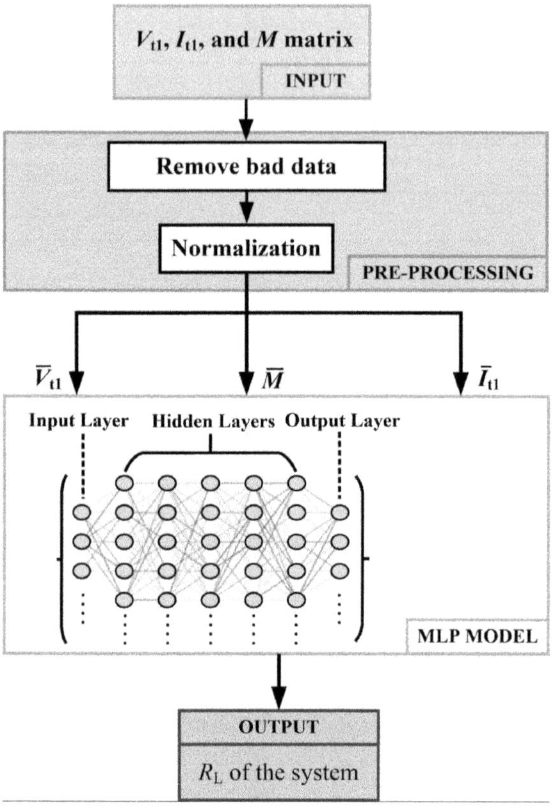

FIGURE 4.12
Schematic diagram of the MLP-based load monitoring strategy. (Adapted from reference [7].)

TABLE 4.2
Key Parameters of the WPT System Used to Validate the MLP-Based Method

Parameter	Value	Parameter	Value
Transmitter inductance L_t	60.9 μH	Receiver inductance L_r	60.8 μH
Transmitter capacitance C_t	42 nF	Receiver capacitance C_r	41.59 nF
Transmitter resistance R_t	0.14 Ω	Receiver resistance R_r	0.14 Ω
Input DC voltage V_{DC}	10 V	Mutual inductance M	15.38 μH

dataset combinations are used, each containing a total of 1000 data points but with varying ratios of simulation to experimental data. Dataset-1 consists of 1000 simulation data points and no experimental data. Dataset-2 includes 960 simulation data points and 40 experimental data points. Dataset-3 contains 940 simulation data points and 60 experimental data points. Dataset-4 comprises 920 simulation data points and 80 experimental data points. Dataset-5 has 900 simulation data points and 100 experimental data points. The training data is split into training and validation subsets, with 80% (800 data points) allocated for training and 20% (200 data points) for validation. An additional 50 experimental data points are reserved as the test set for final evaluation. TensorFlow 2 is utilized as the ML framework due to its ease of deployment in production environments. The model achieves a prediction execution time of 0.047 seconds and occupies 1.7 GB of memory.

The loss functions for both training and testing across the five cases are illustrated in Figure 4.13. The mean absolute errors (MAEs) for both training and test losses demonstrate convergence, indicating that the MLP models are well-trained without overfitting.

Figure 4.14 compares the load estimation errors (relative errors) between the proposed strategy, trained on five datasets, and the conventional model-based method. In Figure 4.14a, the proposed strategy using MLP trained on Dataset-1 and Dataset-2 is evaluated, showing average relative errors of 257% and 28.88%, respectively. The MLP trained on Dataset-2 significantly reduces estimation errors compared to the MLP trained solely on Dataset-1 (pure simulation data). In Figure 4.14b, the proposed strategy using MLP trained on Dataset-2, Dataset-3, Dataset-4, and Dataset-5 is compared with the conventional model-based method. Figure 4.14c summarizes the average relative errors across these datasets and methods, with errors of 23.59%, 17.94%, 10.78%, and 22.72% for Dataset-3, Dataset-4, Dataset-5, and the conventional method, respectively. As the proportion of experimental data in the training set increases, the relative errors decrease. Notably, the proposed method using Dataset-4 and Dataset-5 achieves higher accuracy than the conventional model-based approach. For applications with specific accuracy and measurement constraints, the ratio of simulation to experimental data in the hybrid dataset can be adjusted to meet the required performance.

4.4.1.3 Discussion on the MLP-Based Method

The MLP-based monitoring method presented in [7] explores the use of AI technologies to model practical parameters that are difficult to describe with simple formulas. These include parameter drifts caused by temperature changes, variable parasitic effects due to fluctuating currents or voltages, and unexpected harmonics. Our studies show that while the MLP-based monitoring strategy performs well at the resonant frequency, its accuracy declines when both mutual coupling and load conditions are monitored. The method provides acceptable accuracy for load monitoring when the mutual inductance or coupling coefficient is known. In fact, the coupling coefficient is a more suitable input parameter since its value remains between 0 and 1, potentially eliminating the need for data normalization. Regarding the hybrid data approach, the fundamental idea is to use simulation data to represent an ideal system while incorporating experimental data to account for uncertainties. Fortunately, this approach has been effective for the WPT system we investigated, and I am confident that alternative AI algorithms could

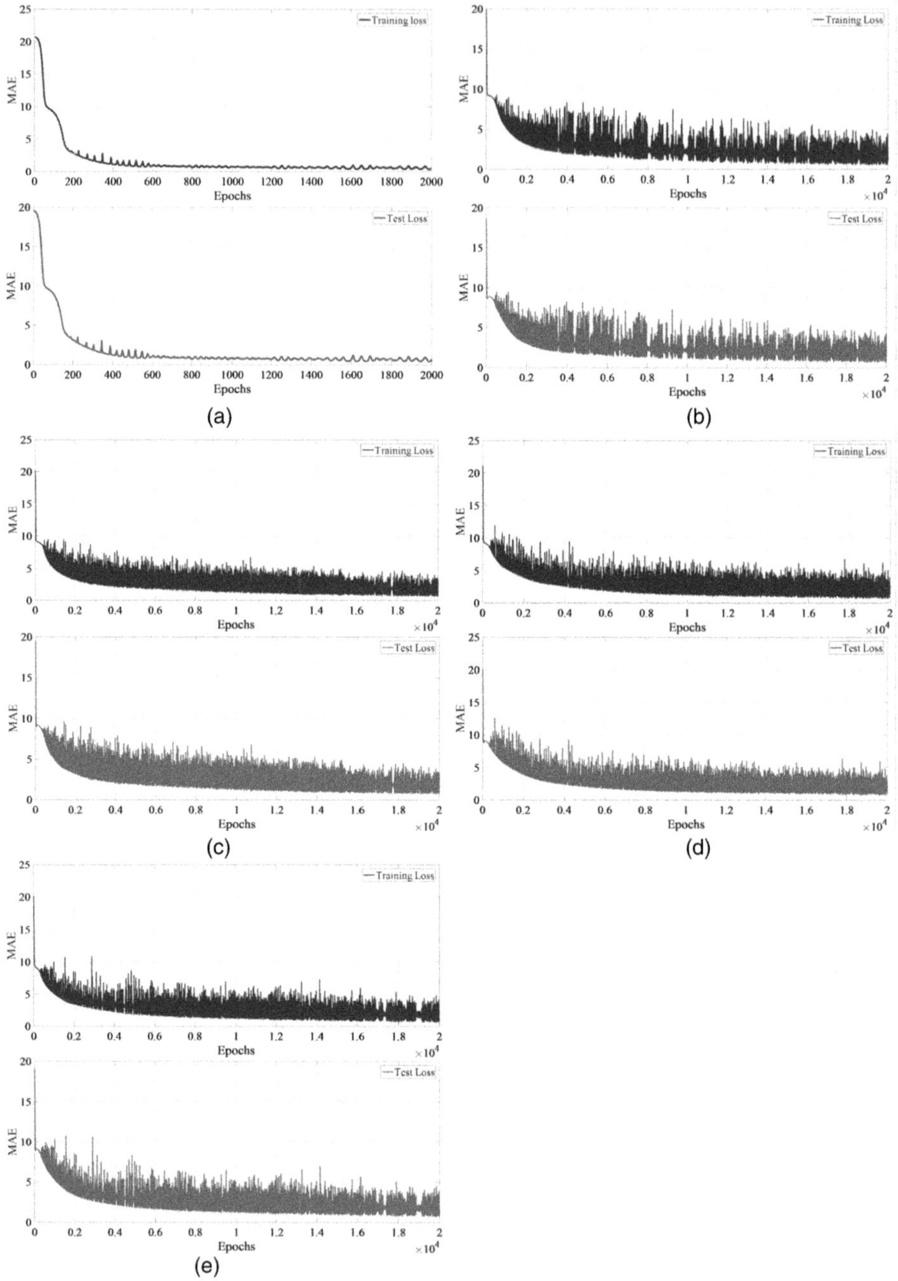

FIGURE 4.13
Plots of training loss and test loss for (a) Dataset-1, (b) Dataset-2, (c) Dataset-3, (d) Dataset-4, and (e) Dataset-5, in experiment. (Adapted from reference [7].)

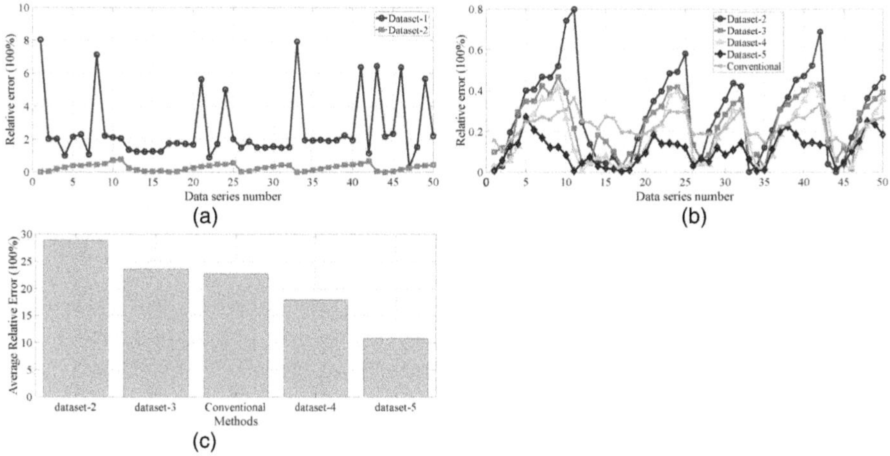

FIGURE 4.14
Comparison of load estimation errors among the MLP-based method with different datasets and the conventional model-based method: (a) MLP-based method with Dataset-1 and Dataset-2, (b) MLP-based method with Dataset-2, Dataset-3, Dataset-4, Dataset-5, and conventional model-based method, and (c) average relative errors. (Adapted from reference [7].)

yield similar results. However, I must admit that the underlying principle remains unclear, and I cannot confirm whether the same logic would apply to other WPT or power electronics systems.

4.4.1.4 Overview of the RF-Based Method

A more specific name of the RF-based method in this book is the Random Forest Regressor (RFR)-based method. The RFR is a meta-estimator that constructs multiple decision trees on various sub-samples of the dataset and uses averaging to enhance predictive accuracy while reducing overfitting [16]. It is based on the Classification and Regression Tree (CART) algorithm, which utilizes a binary tree structure to perform linear regression tasks. When the output values are continuous variables, CART is well-suited for making predictions. RF is a widely used ML technique that requires fewer data samples compared to deep learning (DL) methods [17]. For online monitoring of mutual inductances of the WPT systems in [8], the available datasets based on the transmitter-side measurements are often limited. Therefore, RF is a more suitable choice than DL.

The classic CART algorithm aims to group data with similar characteristics into the same class while maximizing the separation between different classes and minimizing variability within each class. The regression algorithm follows a similar principle, except that instead of classifying data, it fits the model to labeled continuous values. Given a dataset:

$$D=\{(x_1,y_1),(x_2,y_2),\ldots,(x_n,y_n),\ldots(x_N,y_N)\} \tag{4.43}$$

Let $x_n = (x_{n,1}, x_{n,2}, \ldots, x_{n,k}, \ldots, x_{n,K})$ represent the input data for a given task, where x_n is the feature set of the n-th sample and a K-dimensional vector. The corresponding output value for the n-th sample is denoted as y_n. The objective of the regression task is to construct a model $f(x_n)$ that predicts the output value with minimal error. RFR is a powerful ensemble learning algorithm composed of multiple independent decision trees acting as weak regressors. The final output of RFR is determined by aggregating the predictions from these weak classifiers. The flowchart of the adopted RFR method is plotted, as shown in Figure 4.15.

Initially, fundamental and harmonic components of v_t and i_t (i.e., v_{t1}, i_{t1}, v_{t3}, i_{t3},...), along with mutual inductance (i.e., M), are measured to construct the dataset. Here, only the root-mean-square (RMS) values V_{ti} and

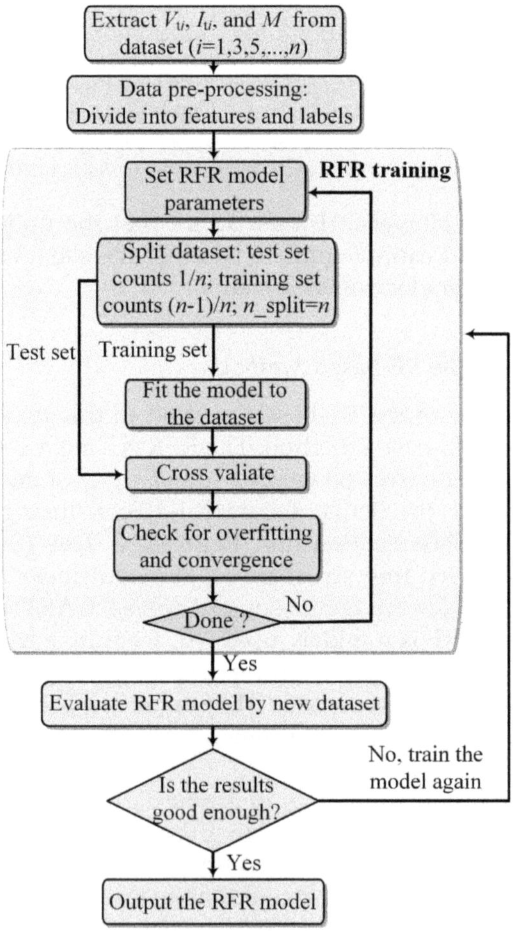

FIGURE 4.15
Flowchart of the adopted RFR algorithm. (Adapted from reference [8].)

I_{ti} are required, while phase angles are not necessary. The collected data undergoes preprocessing, where it is divided into features and labels. The key parameters of the RFR algorithm are listed in Table 4.3. The dataset is then split into a training set and a test set. The training set is used to fit the RFR model, while the test set is used for cross-validation. If overfitting occurs, the model parameters are adjusted. If the model converges successfully, it is further evaluated using a new dataset with additional measurements under different operating conditions. If the RFR model performs well across all datasets, it is finalized. Otherwise, further training is required.

The schematic diagram of the proposed RFR-based monitoring scheme is depicted in Figure 4.16. The input parameter matrix consists of V_{t1}, V_{t3}, V_{t5}, V_{t7}, I_{t1}, I_{t3}, and I_{t5}, while higher-order harmonics are considered negligible. After filtering out erroneous data, the remaining dataset is normalized and used as input for the RFR model. The model then outputs the estimated mutual inductance (i.e., $\tilde{M}$). Notably, the input voltage and current measurements are taken at resonant frequencies, enabling real-time online monitoring during the power transfer process.

TABLE 4.3

Key Parameters of the Adopted RFR Algorithm

Parameter	Meaning
n_estimators	Number of trees in the forest
criterion	Function to measure the quality of a split
max_depth	Maximum depth of each tree
min_samples_split	Minimum number of samples required to split an internal node
min_samples_leaf	Minimum number of samples required at a leaf node
min_weight_fraction_leaf	Minimum weighted fraction of total weights required at a leaf node
max_features	Number of features considered for the best split
max_leaf_nodes	Maximum number of leaf nodes, grown in best-first order
min_impurity_decrease	Threshold for node splitting based on impurity reduction
Bootstrap	Whether bootstrap samples are used for building trees
oob_score	Whether out-of-bag samples are used for estimating generalization error
n_jobs	Number of parallel jobs for computation
random_state	Controls the randomness of both bootstrapping (if bootstrap=True) and feature sampling (if max_features<n_features)
verbose	Controls verbosity during fitting and prediction
warm_start	Whether to reuse previous solutions to add more estimators
ccp_alpha	Complexity parameter for minimal cost-complexity pruning
max_samples	Number of samples used to train each base estimator
n_split	Number of cross-validation splits
Shuffle	Whether to shuffle arrays or sparse matrices consistently

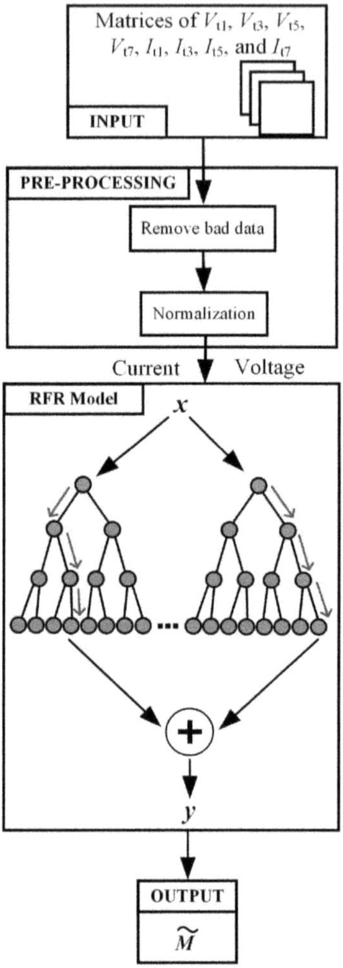

FIGURE 4.16
Schematic diagram of the proposed monitoring scheme based on the RFR. (Adapted from reference [8].)

4.4.1.5 Experimental Verification of the RF-Based Method

Experiments are conducted on an SS-compensated WPT system with the key parameters given in Table 4.2. A total of 1000 experimental data points (i.e., $n=1000$) under varying load conditions are collected to train the RFR model, while an additional 600 experimental data points (i.e., $n=600$) are gathered for comparison with the conventional method. The training and testing of the RFR model are performed using PyCharm 2021, with the default training parameters detailed in Table 4.4. The quality of the RFR model (i.e., Q) is evaluated based on the root mean square error (RMSE) of the estimated mutual inductances relative to the average mutual inductance (i.e., $\overline{M}$), as given by:

TABLE 4.4

Key Parameter Values of the RF-Based Model

Parameter	Value	Parameter	Value
n_estimators	100	criterion	'Squared_error'
max_depth	None	min_samples_split	2
min_samples_leaf	1	min_weight_fraction_leaf	0
max_features	'Auto'	max_leaf_nodes	None
min_impurity_decrease	0	bootstrap	True
bootstrap	False	n_jobs	None
random_state	None	verbose	0
warm_start	False	ccp_alpha	0
max_samples	None	n_split	5
shuffle	True		

$$Q = \frac{\text{RMSE}}{\bar{M}} \tag{4.44}$$

$$\text{RMSE} = \sqrt{\frac{\sum_{i=1}^{n}\left(\tilde{M} - M\right)^2}{n}} \tag{4.45}$$

Here, a smaller Q value indicates a better RFR model.

Initially, the RFR model is trained using default parameters. The dataset is split into a training set comprising 80% of the data (800 samples) and a test set comprising the remaining 20% (200 samples). During training, the back-propagation algorithm is employed to optimize the model, ensuring that the training error remains lower than the test error. For the training set, the RMSE of the estimated mutual inductances is 3.798×10^{-7}, resulting in a Q value of 2.83%, which is within the acceptable threshold of 5%. For the test set, the RMSE is 9.834×10^{-7}, yielding a Q value of 7.33%, which is below the 10% threshold. The weights assigned to the fundamental and harmonic components of V_t and I_t are summarized in Table 4.5, indicating that higher-order currents and voltages contribute minimally to the accuracy of mutual inductance estimation. To further refine the model, the fifth- and seventh-order currents (I_{p5} and I_{p7}) and voltages (V_{p5} and V_{p7}) are excluded from the training dataset. As a result, the Q values increase for both the training and test sets. Specifically, the Q value for the training set rises to 3.01% (RMSE=4.041×10^{-7}), while the Q value for the test set increases to 7.66% (RMSE=1.028×10^{-6}). The weights of the fundamental and third-order harmonics are presented in Table 4.6. These results highlight that incorporating higher-order harmonic components enhances the accuracy of the RFR model.

Next, the parameters n_split, n_estimator, max_depth, min_samples_split, min_samples_leaf, and max_features of the RFR model are fine-tuned using different values. The corresponding Q values for various RFR parameter

TABLE 4.5

Weights of Input Measurements before Removal

Input Parameter	Weight	Input Parameter	Weight
I_{t1}	0.643924	I_{t3}	0.220133
I_{t5}	0.085591	I_{t7}	0.049829
V_{t1}	0.000500	V_{t3}	0.000023
V_{t5}	0.000000	V_{t7}	0.000000

TABLE 4.6

Weights of Input Measurements after Removal

Input Parameter	Weight	Input Parameter	Weight
I_{t1}	0.746717	I_{t3}	0.245562
V_{t1}	0.003290	V_{t3}	0.004431

TABLE 4.7

Q Values of the Training Set and Test Set with Different RFR Parameters

Parameter	Training Set (%)	Test Set (%)
n_split=5	2.83	7.45
n_split=8	2.85	7.21
n_split=10	2.79	7.21
n_estimator=200	2.77	7.16
n_estimator=20	3.13	7.77
max_depth=35	2.98	7.66
min_samples_split=5	3.55	7.78
min_samples_leaf=2	4.05	7.63
max_features='sqrt'	2.93	7.82
max_features='log2'	2.65	7.28

settings are presented in Table 4.7. For all configurations, the Q values remain below 5% for the training set and 10% for the test set, confirming that the trained RFR model achieves a satisfactory level of accuracy.

The additional 600 experimental data points acquired at the resonant frequency are used as inputs for the RFR model to estimate mutual inductances. The estimation errors of the proposed RFR-based monitoring scheme are illustrated in Figure 4.17, with an average estimation error of 2.55% and a standard deviation of 4.04%. For comparison, the same 600 data points are used as inputs for the conventional circuit-based method to estimate mutual inductance. The average estimation error is 5.09%, nearly double that of the RF-based method, with a higher standard deviation of 4.93%. These results confirm that the RF-based method offers greater accuracy and stability compared to the conventional circuit-based approach.

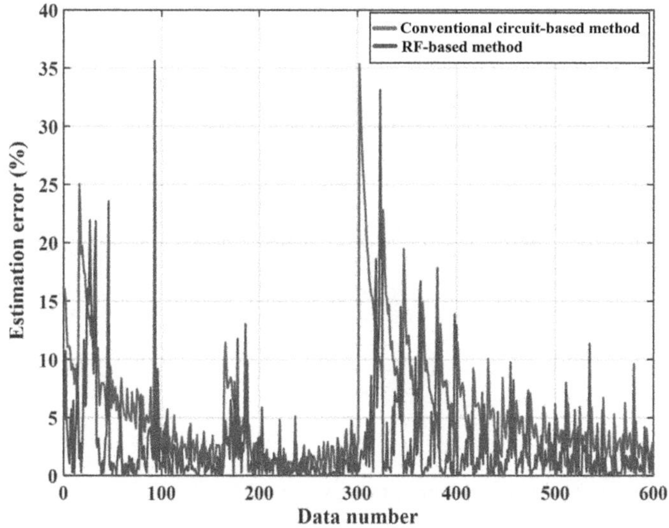

FIGURE 4.17
Estimation errors of the RF-based method and conventional circuit-based method. (Adapted from reference [8].)

4.4.1.6 Discussion on the RF-Based Method

By leveraging the harmonic components of both the transmitter-side input voltage and current, the mutual inductance of WPT systems can be monitored without the need for communication feedback from the receiver side, including load conditions. The proposed method accounts for the high costs associated with voltage and current sensors required for measuring phase angles of high-order harmonics. Therefore, only the amplitudes or RMS values of these components are used as inputs. The proposed RF-based monitoring strategy offers several advantages over conventional circuit-based approaches. First, it has enhanced robustness against disturbance and parameter drifts. Disturbances in WPT systems may arise from factors such as foreign objects, electromagnetic interference (EMI), and cyberattacks. Additionally, parameter drifts can be caused by current stress, thermal effects, component aging, and other environmental influences. Second, it has faster monitoring at the resonant frequency instead of sub-resonant frequencies. The power transfer efficiency of a WPT system at its resonant frequency is typically higher than at sub-resonant frequencies, excluding converter losses. As a result, many WPT products have standardized resonant frequencies as their fixed operating points. Third, it is easier to implement. Conventional circuit-based monitoring schemes rely on additional circuits such as power supplies, high-performance comparators, wide-bandwidth optical amplifiers, and Schmitt triggers that are necessary. However, the proposed monitoring scheme eliminates the need for those hardware components.

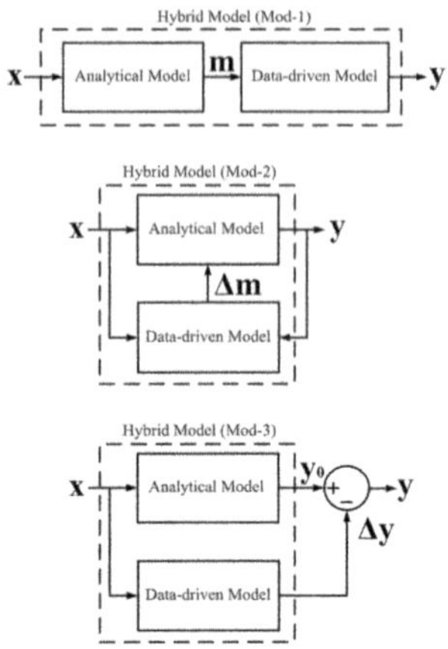

FIGURE 4.18
Schematic diagrams of basic hybrid monitoring schemes.

4.4.2 Hybrid Scheme

Hybrid schemes combine analytical monitoring approaches, such as dynamic response-based and harmonic-based methods, with data-driven models. Figure 4.18 exhibits the schematic diagrams of typical hybrid schemes used for transmitter-side monitoring of critical receiver-side parameters in WPT systems. Three basic structures are plotted for illustration.

In the first structure (Mod-1), the analytical models and data-driven models are sequential. For instance, the intermediate outputs of the analytical model are the inputs of the data-driven model, or vice versa. They can also be cascaded structure, such as an analytical model followed by a data-driven model, which is further followed by an analytical model. The analytical models here could be the models of coupled resonators and the battery load, while the data-driven model is used to describe the rectifier and the filter capacitor, which may comprise a lot of parasitic components. The data-driven model here can also help describe some uncertainties that have not been fully considered in the neighboring analytical models. In the second structure (Mod-2), the analytical and data-driven models follow a "master-slave" configuration. The data-driven model adjusts specific intermediate parameters based on input-output feedback. While the dataset size may be limited, it should focus on key parameters that are challenging to determine accurately. In the third structure (Mod-3), the analytical and data-driven models operate in a "partnership" configuration, running in parallel. The outputs of the data-driven model complement those of the analytical model, refining and enhancing the final results.

If a purely analytical model is considered a "white model" and a purely data-driven model a "black model", hybrid models can be more precisely classified along a spectrum between these two extremes, rather than being broadly labeled as "gray models". The schematic diagram of the model spectrum is depicted, as shown in Figure 4.19. The hybrid model appears "whiter"

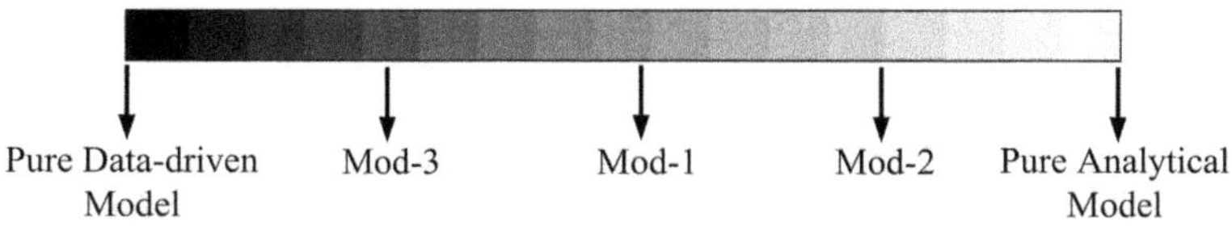

FIGURE 4.19
Schematic diagram of the model spectrum.

when the analytical components are more dominant, whereas it becomes "darker" as the data-driven components take precedence. In practice, balancing the proportions of the analytical and data-driven components is crucial for ensuring accurate monitoring.

References

1. Qi Specification [Online]: https://www.wirelesspowerconsortium.com/knowledge-base/specifications/download-the-qi-specifications/.
2. Y. Yang, J. Zeng, and S. Y. R. Hui, "A fast primary-side current and voltage control for direct wireless battery chargers," *IEEE J. Emerg. Sel. Topics in Power Electron.*, vol. 12, no. 1, pp. 355–366, Feb. 2024.
3. The Qi Wireless Power Transfer System Power Class 0 Specifications: Part 4: Reference Designs, Version 1.2.4, Feb. 2018, p. 7 [Online]: https://www.wirelesspowerconsortium.com/.
4. Y. Yang, "Precise modeling of nonlinear rectifier loads in wireless power transfer systems," *IEEE J. Emerg. Sel. Topics in Power Electron.*, vol. 11, no. 3, pp. 3574–3585, Jun. 2023.
5. Z. Sun, J. Zeng, and Y. Yang, "A resonant-frequency-based dynamic front-end monitoring strategy for series-series-compensated inductive power transfer systems without communication," *IEEE Trans. Ind. Appl.*, in press, 2025.
6. J. Zeng, K. Li, H. Yuan, Y. Yang, S. C. Tan, T. Parisini, and S. Y. R. Hui, "A fast front-end monitoring method for mutual inductance and load movement in SS-compensated wireless power transfer systems based on harmonics extraction," *IEEE Trans. Power Electron.*, vol. 40, no. 1, pp. 2569–2580, Jan. 2025.
7. Y. Yang and H. Wu, "A robust primary-side hybrid data-driven load monitoring strategy for wireless power transfer systems," in *IEEE Energy Conversion Congress and Exposition (ECCE)*, Detroit, MI, USA, 2022, pp. 1–6.
8. Y. Yang, X. Zhang, and K. Wang, "Robust random forest regressor-based mutual inductance monitoring scheme for inductive power transfer systems," in *IEEE International Power Electronics and Application Conference and Exposition (PEAC)*, Guangzhou, Guangdong, China, 2022, pp. 171–176.
9. X. Li, Z. Sun, K. Chen, and Y. Yang, "Artificial intelligence assisted adaptive controller design with position robustness in a GaN-based wireless power transfer system," in *IEEE 7th International Electrical and Energy Conference (CIEEC)*, Harbin, China, 2024, pp. 3444–3450.

10. X. Li, Z. Sun, X. Zhang, Y. Jiang, K. Mao, and Y. Yang, "A robust artificial intelligence-empowered adaptive proportional-integral control for wireless power transfer systems," *IEEE Trans. Ind. Appl.*, doi:10.1109/TIA.2025.3572075.
11. V. K. Ojha, A. Abraham, and V. Snášel, "Metaheuristic design of feedforward neural networks: A review of two decades of research," *Engineering Applications of Artificial Intelligence*, vol. 60, pp. 97–116,2017.
12. X. Glorot, A. Bordes, and Y. Bengio, "Deep sparse rectifier neural networks," in *Proceedings of the Fourteenth International Conference on Artificial Intelligence and Statistics*, Fort Lauderdale, FL, USA, 2011, vol. 15, pp. 315–323.
13. K. Jarrett, K. Kavukcuoglu, M. Ranzato, and Y. LeCun, "What is the best multi-stage architecture for object recognition?," in *2009 IEEE 12th International Conference on Computer Vision*, Kyoto, Japan, 2009, pp. 2146–2153.
14. I. Goodfellow, Y. Bengio, and A. Courville, *Deep Learning*. Cambridge, MA: MIT Press, 2016.
15. T. Guillod, P. Papamanolis, and J. W. Kolar, "Artificial neural network (ANN) based fast and accurate inductor modeling and design," *IEEE Open J. Power Electron.*, vol. 1, pp. 284–299, Jul. 2020.
16. L. Breiman, "Random forests," *Machine Learning*, vol. 45, pp. 5–32, Apr. 2001.
17. E. Alpaydin, *Machine Learning*. Cambridge, MA: The MIT Press, 2021.

5

Transmitter-Side Control for the Next-Generation WPT Systems

5.1 Dependent Transmitter-Side Control

In this book, dependent transmitter-side control refers to control strategies that utilize the transmitter-side monitoring as a substitute for communication feedback from the receiver to the transmitter. Here, communication feedback specifically refers to out-of-band communication, which relies on external communication systems such as Near-Field Communication (NFC), Bluetooth, Wi-Fi, Zigbee, Infrared (IR) Communication, Ultra-Wideband (UWB), and Radio-Frequency Identification (RFID). However, in-band communication techniques, such as ASK and FSK, are not included in this classification.

5.1.1 Dependent Transmitter-Side Control at Resonant Frequency

5.1.1.1 Overview of the Control Strategy

A typical dependent transmitter-side control scheme at resonant frequency is depicted, as shown in Figure 5.1. The resonant-frequency monitoring methods discussed in Chapter 4 can be utilized to directly or indirectly monitor receiver-side parameters, providing feedback for the transmitter-side controller.

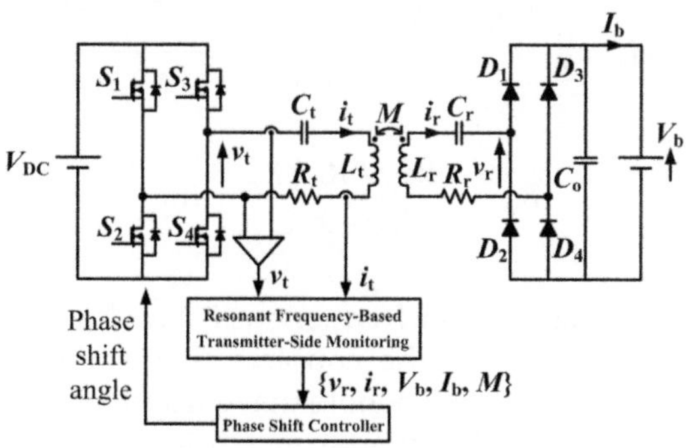

FIGURE 5.1
Schematic diagram of a typical transmitter-side control with resonant frequency-based monitoring schemes.

DOI: 10.1201/9781003468523-5

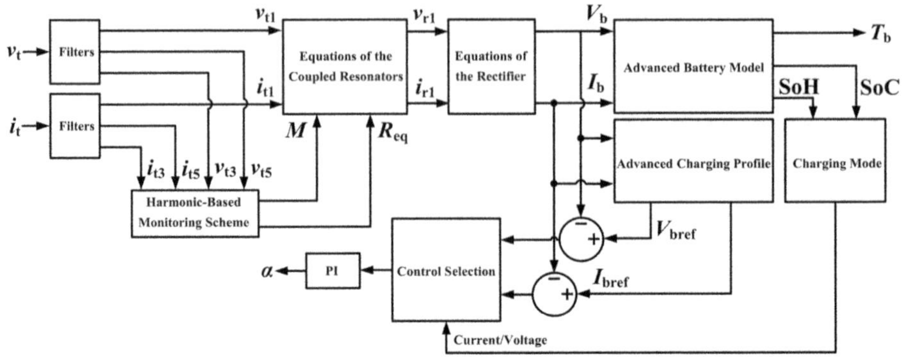

FIGURE 5.2
Control block diagram of the exampled transmitter-side control at resonant frequency.

Given the system's operation at the resonant frequency, phase shift control is a commonly used strategy for the full-bridge inverter.

Here, I provide a more detailed example to illustrate this type of dependent transmitter-side control in WPT systems operating at the resonant frequency. The control block diagram is shown in Figure 5.2. Initially, the input voltage and current of the transmitter (i.e., v_t and i_t) are measured using high-precision voltage and current sensors to capture as many data points as possible with high accuracy. Next, the fundamental and harmonic components of v_t and i_t are separated using bandpass filters, which can be implemented either through software-based methods (such as FFT) or hardware-based solutions (such as analog filter circuits). The third-order and fifth-order harmonic components are subsequently input into the harmonic-based monitoring scheme to estimate the mutual inductance and load conditions using

$$M_n = \sqrt{\frac{\left[\left(\frac{R_{inn} - R_{tn}}{X_{tn} - X_{inn}}\right)^2 + 1\right] X_{rn} (X_{tn} - X_{inn})}{n^2 \omega^2}} \tag{5.1}$$

$$R_{eq} = \frac{(R_{inn} - R_{tn}) X_{rn}}{X_{tn} - X_{inn}} - R_{rn} \tag{5.2}$$

where ω is the switching angular frequency. R_{inn} and X_{inn} represent the real and imaginary components of the coupled resonator's input impedance, respectively. Similarly, R_{tn} and X_{tn} denote the real and imaginary components of the transmitting resonator's impedance, while R_{rn} and X_{rn} correspond to the real and imaginary components of the receiving resonator's impedance. n indicates the number of harmonics. R_{eq} represents the equivalent load resistance of the nonlinear rectifier load. Meanwhile, the fundamental components are utilized to estimate the output voltage and current of the receiving resonator using

$$i_{r1} = \frac{Z_t i_{t1} - v_{t1}}{j\omega M} \tag{5.3}$$

$$v_{r1} = R_{eq} i_{r1} \tag{5.4}$$

where Z_t is the transmitting resonator's impedance. Then, the output charging current and voltage can be estimated using

$$I_b = \frac{2}{\pi} I_{r1\_peak} \tag{5.5}$$

$$V_b = \frac{\pi}{4} V_{r1\_peak} \tag{5.6}$$

where I_{r1_peak} and V_{r1_peak} are the peak values of i_{r1} and v_{r1}. Similar to the bandpass filters, peak detection can be implemented either through software-based methods (e.g., FFT) or hardware-based solutions (e.g., analog filter circuits). The estimated V_b and I_b can be further utilized to determine the battery temperature (i.e., T_b), SoC, and SoH by using an advanced battery model [1]. These derived SoC and SoH can then be used to determine whether the battery should be charged in current mode or voltage mode [2]. The charging current reference (i.e., I_{bref}) and charging voltage reference (i.e., V_{bref}) can be determined based on advanced charging profiles, as demonstrated in our previous studies [1–4]. The charging current and charging voltage are regulated to closely follow their respective reference values. In this case, a conventional proportional-integral (PI) controller is employed. The output of the PI controller corresponds to the phase shift angle α of the transmitter-side inverter. Figure 5.3 exhibits the schematic diagram of the switching signals, the corresponding output voltage of the inverter, and the phase shift angle α. S_1 and S_2 are complementary, as are S_3 and S_4, to prevent short-circuiting of the bridge legs. The phase shift occurs between S_1 and S_3, as well as between S_2 and S_4. During these phase shift intervals, the input DC voltage is short-circuited, causing the inverter's output voltage v_t to drop to zero. By adjusting the phase shift angle α, the amplitude of the fundamental voltage v_{t1} can be effectively regulated. Defining output boundaries for the phase shift angle is unnecessary, as any phase shift angle can be equivalently mapped to a value within the range of 0°–180°. For instance, if the value of α calculated by the PI controller is 765°, the inverter's performance remains identical to that when $\alpha = 45°$. Additionally, as the defined phase difference

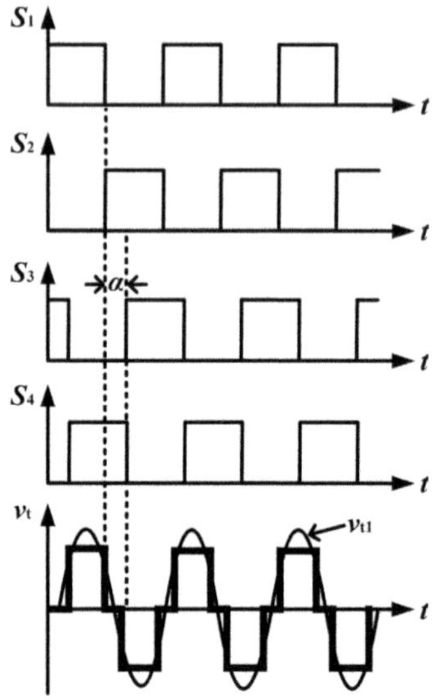

FIGURE 5.3
Schematic diagram of the output voltage and current, the phase shift angle of the transmitter-side inverter, and the switching signals.

TABLE 5.1

Main Parameters of the WPT System Used to Validate the Dependent Transmitter-Side Control at Resonant Frequency

Parameter	Values
Transmitter and receiver inductances L_t, L_r	49.2 μH
Transmitter and receiver capacitance C_t, C_r	51.48 nF
Transmitter and receiver resistance R_t, R_r	0.283 Ω
Input DC voltage V_{DC}	10 V
Sampling rate of the controller	1×10^{-7}s
Tuning coefficient of the proportional gain K_p	–1
Tuning coefficient of the integral gain K_i	–50,000

increases, the output current or voltage decreases. Consequently, if the input to the PI controller is the difference between the reference and the output, the tuning coefficients of the PI controller will be negative.

5.1.1.2 Simulation Results

Simulations are carried out using MATLAB/Simulink to validate the effectiveness of the proposed dependent transmitter-side control strategy at resonant frequency. The study focuses on an SS-compensated WPT system. The main parameters of the system, along with the tuning parameters of the PI controller, are detailed in Table 5.1. The system's operating frequency

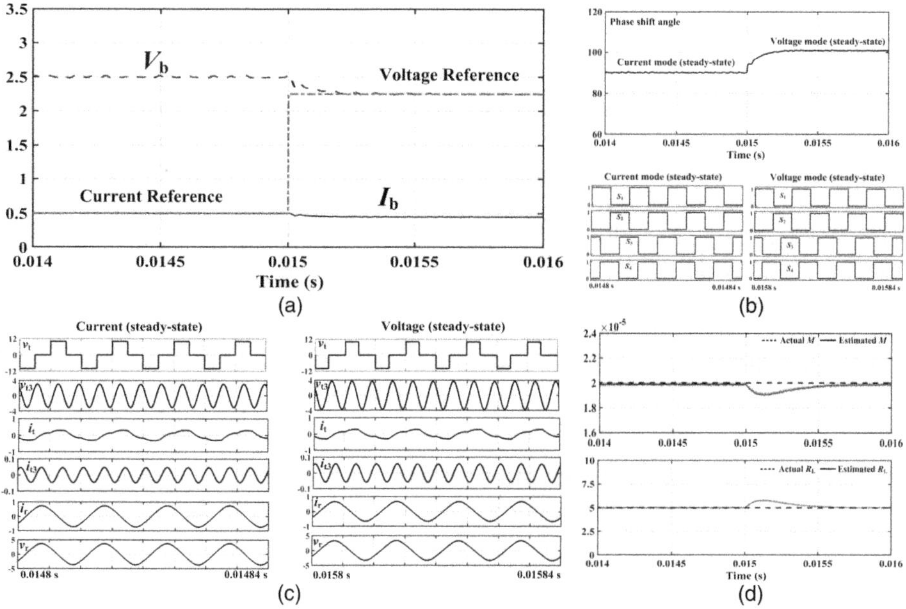

FIGURE 5.4
Measured waveforms of the SS-compensated WPT system under transmitter-side control at the resonant frequency: (a) charging current I_b and charging voltage V_b, (b) phase shift angle u and gate switching signals, (c) key electrical parameters of the coupled resonators, and (d) estimated mutual inductance M and load resistance R_L.

and the resonant frequencies of the coupled resonators are both 100 kHz. Figure 5.4a shows the waveforms of I_b and V_b as they track their respective references (i.e., I_{bref} and V_{bref}) during the transition from current mode to voltage mode. The corresponding waveforms of the switching signals (i.e., S_1, S_2, S_3, and S_4), the phase shift angles, as well as the voltages and currents of the resonators are depicted in Figure 5.4b and c, respectively. The estimated mutual inductance (i.e., M) and equivalent load resistance (i.e., R_L) are shown in Figure 5.4d. Clearly, I_b and V_b track their references with high accuracy. The corresponding phase shift angles are 89.9° and 100.8°, respectively. Additionally, the estimated values of mutual inductance M and equivalent resistance R_{eq} closely match the actual values, with relative estimation errors of 1.1% and 1.8%, respectively.

This example demonstrates the effectiveness of employing a dependent transmitter-side control strategy to charge a battery load in a typical WPT system operating at the resonant frequency. Clearly, the harmonic-based monitoring scheme could be replaced by other resonant frequency-based methods (such as a dynamic response-based scheme), while the core control remains based on conventional phase shift control. Although phase shift control might be substituted with advanced fixed-frequency techniques like pulse density modulation, the novelty of the dependent transmitter-side control does not lie in the control algorithm alone. In fact, improvements are limited by simply changing the control algorithm (e.g., from PI to nonlinear control). The advanced performance of the dependent transmitter-side control mainly hinges on the accuracy and computational efficiency of the transmitter-side monitoring, as well as other factors like the battery model performance and the charging profile. Therefore, for emerging WPT systems that require continuous transmitter-side control at the resonant frequency, control designers should consider adopting the dependent transmitter-side control strategy. However, the primary design focus must be on achieving high model accuracy for both the WPT system and the battery load, ensuring the security, reliability, and longevity of the charging profile, and balancing the trade-offs between hardware sensing circuits and software signal processing.

5.1.2 Dependent Transmitter-Side Control at Non-Resonant Frequency

5.1.2.1 Overview of the Control Strategy

Figure 5.5 shows a typical dependent transmitter-side control at non-resonant frequency. The non-resonant frequency-based monitoring methods discussed in Chapter 4 can be employed to directly or indirectly monitor receiver-side parameters, serving as feedback for the transmitter-side controller. Since the system operates at a non-resonant frequency, frequency control is a commonly adopted strategy for the full-bridge inverter.

Similarly to the resonant frequency case, I present an example to illustrate the design and performance of a dependent transmitter-side control at the

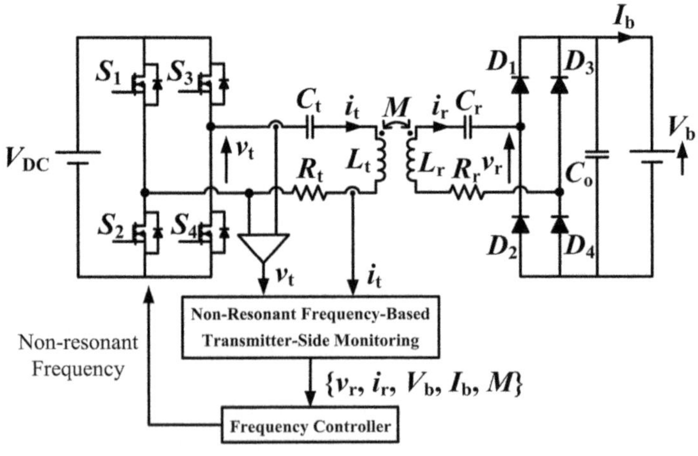

FIGURE 5.5
Schematic diagram of a typical transmitter-side control with non-resonant frequency-based monitoring schemes.

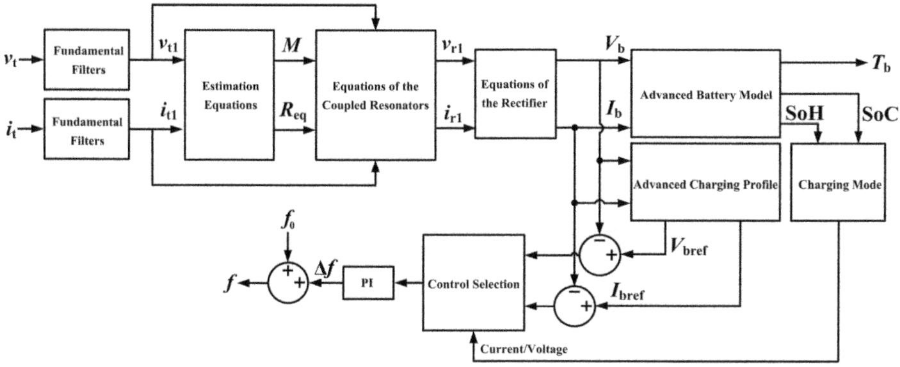

FIGURE 5.6
Control block diagram of the exampled transmitter-side control at non-resonant frequency.

non-resonant frequency. Figure 5.6 presents the control block diagram for the studied dependent transmitter-side control operating at a non-resonant frequency. Initially, the fundamental components of v_t and i_t (i.e., v_{t1} and i_{t1}) are measured using bandpass filters. Based on the estimation equations for the non-resonant frequency [5], the mutual inductance (i.e., M) and load resistance (i.e., R_{eq}) can be determined using

$$M = \sqrt{\frac{\omega L_r - \dfrac{1}{\omega C_r}}{\omega^2 \mathbf{Im}\left(Z_{in} - Z_t\right)^{-1}}} \tag{5.7}$$

$$R_{eq} = \frac{\omega L_r - \dfrac{1}{\omega C_r}}{\left(Z_{in} - Z_t\right)\mathbf{Im}\left(Z_{in} - Z_t\right)^{-1}} - Z_r \tag{5.8}$$

where $\mathbf{Im}(\cdot)$ indicates the imaginary part of a complex number. Z_{in} is the input impedance of the coupled resonator. Using Equations (5.7) and (5.8), the output voltage and current of the receiving resonator can be estimated from Equations (5.3) and (5.4). Subsequently, the output charging current and voltage can be determined using Equations (5.5) and (5.6). The feedback loops for I_b and V_b, as well as the determination of I_{bref} and V_{bref}, follow the same principles as those in the transmitter-side control scheme at the resonant frequency. The key difference is that the controller output is Δf instead of the phase shift angle α. This frequency deviation Δf is then added to the resonant frequency f_0 to set the switching frequency for the transmitter-side inverter. The relationship between the output current or output voltage and the operating frequency of a typical WPT system is not very straightforward. The system parameters, including the mutual inductance and load conditions, will affect the characteristics of outputs versus the operating frequency.

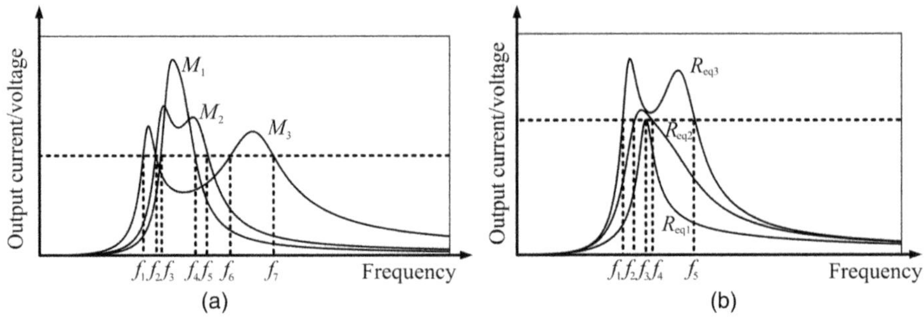

FIGURE 5.7
Schematic waveforms illustrating output variations at different operating frequencies for WPT systems with (a) varying mutual inductance and (b) varying load resistance.

The schematic curves are shown in Figure 5.7. The bifurcation phenomenon occurs when the mutual inductance of the WPT system increases (i.e., $M_3 > M_2 > M_1$) or when the load resistance decreases (i.e., $R_{eq3} < R_{eq2} < R_{eq1}$). As a result, multiple operating frequencies may be possible to achieve the desired output current or voltage. Therefore, the control strategy must be carefully designed to prevent oscillations between feasible operating frequencies. One possible solution to ensure the WPT system operates at a stable, fixed frequency is presented in [6].

5.1.2.2 Simulation Results

Simulations are conducted using MATLAB/Simulink to validate the effectiveness of the proposed dependent transmitter-side control strategy. The main parameters of the system are given in Table 5.1, while the tuning parameters of the PI controller are $K_p = 50$ and $K_i = 10{,}000$. Figure 5.8a presents the waveforms of I_b and V_b as the system transitions from current mode to voltage mode. The corresponding waveforms of the switching signals are presented in Figure 5.8b. Figure 5.8c displays the waveforms of key voltages and currents of the resonators, while Figure 5.8d presents the estimated mutual inductance and equivalent load resistance. Apparently, I_b and V_b accurately track their reference values. The corresponding operating frequencies are 110 and 110.8 kHz, respectively. At the two non-resonant operating frequencies, the mutual inductance M and the equivalent resistance R_{eq} can be accurately monitored. The monitoring errors are 1.8% and 0.4% at 110 kHz, and 1.1% and 2.7% at 110.8 kHz, respectively.

This example demonstrates the efficacy of a dependent transmitter-side control scheme operating at a non-resonant frequency to regulate a battery load in a typical WPT system. Instead of transmitter-side monitoring, alternative approaches, such as heuristic algorithm-based frequency sweeping, auxiliary circuit-based frequency sweeping, or AI-based monitoring (as discussed in the previous chapter), could be employed.

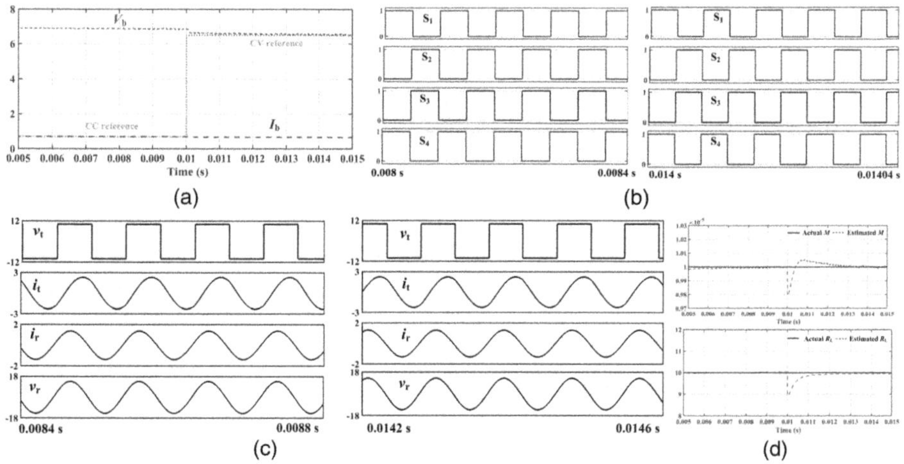

FIGURE 5.8
Measured waveforms of the SS-compensated WPT system under transmitter-side control at non-resonant frequencies: (a) charging current I_b and charging voltage V_b, (b) gate switching signals, (c) key electrical parameters of the coupled resonators, and (d) estimated mutual inductance M and load resistance R_L.

The frequency control method used here is the well-known classical approach, renowned for its superior soft-switching performance in the transmitter-side inverter compared to conventional phase shift control across a wide operating range. Furthermore, this control strategy can be adapted into a variable-frequency phase shift control scheme to expand control flexibility, enabling additional objectives such as output current regulation, zero-voltage switching (ZVS), and maximum efficiency optimization. However, it should be noted that the novelty of dependent transmitter-side control schemes at non-resonant frequencies may stem primarily from transmitter monitoring rather than the controller, similar to the dependent transmitter-side control at resonant frequency. However, frequency-regulated dependent transmitter-side control at non-resonant frequencies possesses unique characteristics not found in phase shift-regulated dependent control at resonant frequencies. For example, the self-oscillating control scheme presented in [6] operates without requiring PI compensators or complex control algorithms typically implemented in digital controllers. Moreover, since dependent transmitter-side control at non-resonant frequencies may offer multiple feasible operating points to achieve the target output current or voltage, the operating frequency can be optimally selected. Frequencies above resonance are often preferred as they facilitate ZVS in the transmitter-side inverter. Among these, a particular frequency may maximize system efficiency for given mutual coupling and load conditions. These considerations do not apply to dependent transmitter-side control schemes at resonant frequency.

5.1.3 Further Discussion on Dependent Transmitter-Side Control

Simulation results demonstrate that accurate transmitter-side monitoring is essential for ensuring the precision of dependent transmitter-side control. Both phase shift and frequency control employ conventional feedback mechanisms, suggesting that the innovative aspects of dependent transmitter-side control are relatively constrained. Consequently, there appears to be limited potential for significant further development of these control strategies. However, the main challenge of dependent transmitter-side control schemes lies in their practicality, as they rely heavily on monitoring accuracy. Many conventional monitoring methods discussed in Chapter 4 are susceptible to disturbances and parasitic parameters, especially those utilizing simple algorithms without additional hardware components. Therefore, in my view, dependent transmitter-side control schemes without incorporating advanced techniques may not meet the requirements of next-generation WPT systems if only simple algorithms are used. That said, this does not render dependent transmitter-side control schemes obsolete. They can still serve as auxiliary methods for detecting significant system changes, such as faults or intrusive foreign objects. Moreover, with the integration of AI technologies mentioned in Chapter 4, mismatches in monitoring strategies can be effectively compensated. This can greatly enhance monitoring accuracy, ultimately enabling precise output control.

Nevertheless, as mentioned in Chapter 1, WPT systems are evolving toward higher power and greater efficiency with advancements in power semiconductors. This raises a critical issue of the power conversions, including the DC-to-AC conversion at the transmitter side and the AC-to-DC conversion at the receiver side, that should not be overlooked. For high-frequency operations in the tens of megahertz, conventional H-bridge topologies may not achieve the desired power rating, even with gallium nitride (GaN) devices. As a result, advanced resonant converters, such as class EF power converters, become necessary. However, most of these converters are either load-dependent or only load-independent within a limited operating range, making output regulation more complex than the straightforward phase shift or frequency control used in typical H-bridge inverters. Therefore, modifications are required for WPT systems utilizing different inverting and rectifying circuits. However, adapting complex model-based monitoring and control schemes (such as heuristic algorithm-based monitoring or dynamic response-based monitoring) may not be straightforward. In contrast, hardware-based frequency sweeping schemes that bypass the load show promise for application in WPT systems with alternative converter topologies. Whether you are a student, an academic researcher, or an electrical engineer, it is worthwhile to explore modifications to the transmitter-side monitoring and control schemes discussed in previous chapters. Adapting these methods for WPT systems with unexplored compensated networks and power converters could lead to valuable advancements in the field.

Beyond the main power converters, it is also crucial to consider the challenges in designing sensors for ultrahigh-frequency WPT systems, especially those measuring AC components, which require extremely high bandwidth. I have three recommendations to address the challenges associated with high-bandwidth sensor requirements in ultrahigh-frequency WPT systems:

i. Prioritize DC-based monitoring and control: Whenever possible, implement monitoring and control strategies that rely on DC components. If AC measurement is unavoidable, consider using analog circuits to convert AC signals into DC signals.
ii. Leverage data-driven approaches: Gather sufficient data to minimize measurement errors using fitting techniques.
iii. Utilize alternative control algorithms: Employ control methods such as self-oscillating control [6], conventional bang-bang control, and data-driven control to reduce reliance on high-bandwidth sensors.

5.2 Independent Transmitter-Side Control

Independent transmitter-side control refers to control strategies that do not rely on the previously mentioned transmitter-side monitoring schemes to achieve communication-free regulation. Here, "communication-free" means that the WPT system operates without external communication, while in-band communication techniques such as ASK and FSK are not considered external communication.

5.2.1 Auxiliary Circuit-Based Control Scheme

5.2.1.1 Overview of System Architecture and Control Strategy

The core concept of the auxiliary circuit-based scheme involves encoding brief but critical information into the power flow, which is then decoded on the transmitter or receiver end of the WPT system. At that time, I was concerned by the limitations of inaccuracies of non-resonant frequency monitoring schemes, as they failed to meet the real-time accuracy requirements for wireless battery chargers. I considered using an innovative approach to convey a simple yet important message without any external communication, much like the beacon towers along the Great Wall. My initial concept involved adding a saturation unit, such as a voltage clip circuit, to the receiver side. When saturation is achieved, the transmitter can detect the receiver's voltage condition, which can then be used to estimate coupling or load conditions. While this method is feasible under resonant frequency, I encountered a critical issue during experimentation: achieving a rapid voltage boost to

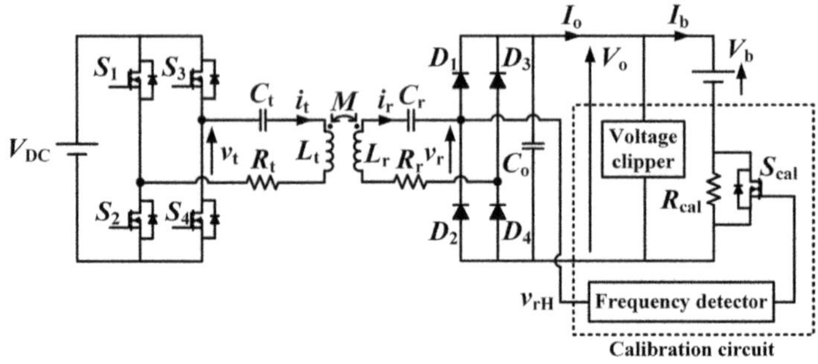

FIGURE 5.9
Schematic diagram of an SS-compensated WPT system with the auxiliary circuit.

saturation is nearly impossible with a battery load, as the battery's terminal voltage cannot change instantaneously. To address this, I considered adding a resistor as a "spring" to quickly elevate the load voltage to saturation. However, the resistor introduces additional power loss. To mitigate this, I proposed incorporating a bypass power switch that short-circuits the resistor during operation while leaving it open during the calibration process. However, this introduced a new challenge: how to effectively control the turn-on and turn-off of the bypass power switch. At the time, I found inspiration in the FSK and ASK techniques used in the "Qi" standard, which led me to develop the auxiliary circuit-based scheme detailed in [7].

The schematic diagram of an SS-compensated WPT system with the auxiliary circuit is depicted, as shown in Figure 5.9. The calibration circuit, consisting of a voltage divider, a calibration switch S_{cal}, a calibration resistor R_{cal}, and a frequency detector, is placed between the diode-bridge rectifier and a battery load. Notably, this calibration circuit is not a switching power regulator. The calibration switch S_{cal} toggles only at the initial moments of the pre-calibration and pre-charging stages, while the calibration resistor R_{cal} remains short-circuited during the long charging period. As a result, integrating the calibration circuit does not impact the EMI or efficiency of the WPT system. The conventional phase shift control method, which is compatible with the "Qi" standard (as demonstrated in reference designs MP-A1, MP-A4, and MP-A7 for extended power transmitters [8]), can be seamlessly applied during both calibration and charging processes. During the calibration stage, the phase shift angle is progressively reduced through open-loop control. In the charging stage, closed-loop phase shift control is employed to regulate both the output current and voltage, ensuring they accurately track their reference values.

5.2.1.2 Rigorous Analysis of System Operations

In the schematic diagram shown in Figure 5.9, the switching signals of each bridge leg are complementary. In conventional phase shift control, the phase shift angle α between the switching signals of the diagonal pairs (i.e., S_1 and S_4, S_2 and S_3) is adjusted to regulate the input voltage of the resonators (i.e., v_t), as depicted in Figure 5.3. The calibration circuit comprises (i) a voltage clipper, (ii) a calibration switch S_{cal}, (iii) a calibration resistor R_{cal}, and (iv) a frequency detector. The voltage clipper is a protection circuit designed to limit V_o to an upper threshold V_{oH} (i.e., $V_o \le V_{oH}$). When V_o reaches V_{oH}, the transmitter current i_t becomes saturated at a predefined value. By measuring this saturated current, the output charging current and charging voltage can be estimated during the calibration stage. The switch S_{cal} remains off during the calibration process, including both the pre-calibration and calibration stages. Once the calibration stage is complete, S_{cal} is turned on by setting $S_{cal}=1$ through the frequency detector. Throughout the entire charging process, including the pre-charging and charging stages, S_{cal} remains on. The resistor R_{cal} is connected in series with the battery load to boost the output voltage V_o to the upper limit V_{oH} during the calibration stage. Since S_{cal} is turned off during calibration, the output voltage V_o is the sum of the battery voltage V_b and the voltage drop across R_{cal}. By reducing the phase shift angle of the transmitter-side inverter, the charging current I_b increases, which in turn raises the voltage across R_{cal}, allowing V_b to reach V_{bH} and causing i_t to saturate. The transmitter-side saturated current is measured to calculate the output charging current I_b and charging voltage V_b, as detailed later. During the charging process, R_{cal} is bypassed by closing S_{cal}, ensuring that the output voltage V_o is equal to the battery voltage V_b and preventing any power dissipation in R_{cal}. The frequency detector identifies the frequency of the nodal voltage v_{rH} (denoted as f_m) and compares it with the reference frequencies f_{mL} and f_{mH}. The high reference frequency f_{mH} is set above the resonant frequency (i.e., $f_{mH} > f_o$), while the low reference frequency f_{mL} is set below it ($f_{mL} < f_o$). If $f_m \ge f_{mH}$, the frequency detector outputs 0 (i.e., $S_{cal}=0$). If $f_m \le f_{mL}$, the output is 1 (i.e., $S_{cal}=1$). However, if $f_{mL} < f_m < f_{mH}$, the output remains unchanged. Further details about the frequency detector will be discussed later.

Table 5.2 summarizes the key differences in the calibration circuit between the calibration and charging processes. Before the transmitter initiates pre-calibration, the inverter's switching frequency is set to be greater than or equal to f_{mH}, ensuring that the frequency detector outputs 0 (i.e., $S_{cal}=0$), keeping R_{cal} in series with the battery load. During the calibration stage, after

TABLE 5.2

Comparison of Calibration Circuit Operation during the Calibration and Charging Processes

	Calibration Process		Charging Process
Topology	Figure 5.10a		Figure 5.10b
Switching condition for S_{cal}	$S_{cal}=0$		$S_{cal}=1$
Output voltage of the rectifier V_o	$V_o=V_b+I_bR_{cal}$		$V_o=V_b$
Detected frequency f_m	$f_m \ge f_{mH}$	$f_{mL} < f_m < f_{mH}$	$f_m \le f_{mL}$

pre-calibration, the inverter operates at the resonant frequency to determine the output charging current I_b and charging voltage V_b. Since the detected frequency f_m falls within the range of f_{mL} and f_{mH}, the frequency detector output remains unchanged (i.e., $S_{cal}=0$). When the calibration stage ends and the pre-charging stage begins, the inverter's switching frequency is adjusted to be less than or equal to f_{mL}, triggering the calibration switch S_{cal} to turn on and short R_{cal} (i.e., $S_{cal}=1$). Throughout the charging stage, the switching frequency is maintained at the resonant frequency f_o, keeping the frequency detector output unchanged (i.e., $S_{cal}=1$) and ensuring that R_{cal} remains shorted. The calibration circuit topologies during the calibration and charging processes are plotted in Figure 5.10. It is worth noting that the voltage clipper and frequency detector functions can also be implemented using existing battery management systems (BMS) and "Qi"-compatible devices with FSK technology. Unlike conventional secondary-side power converters, which utilize high-frequency switching signals to regulate power flow in the WPT systems, the proposed calibration circuit operates for a significantly shorter duration. Rather than functioning as a switched-mode power converter, it effectively acts as an FSK demodulator.

5.2.1.3 Equivalent Circuit Analysis

Based on the schematic circuit diagram in Figure 5.9, the equivalent circuit of the resonators at the fundamental frequency is depicted in Figure 5.11. Here, ω represents the switching angular frequency, while v_{t1}, i_{t1}, i_{r1}, and v_{r1} denote the fundamental components of v_t, i_t, i_r, and v_r, respectively.

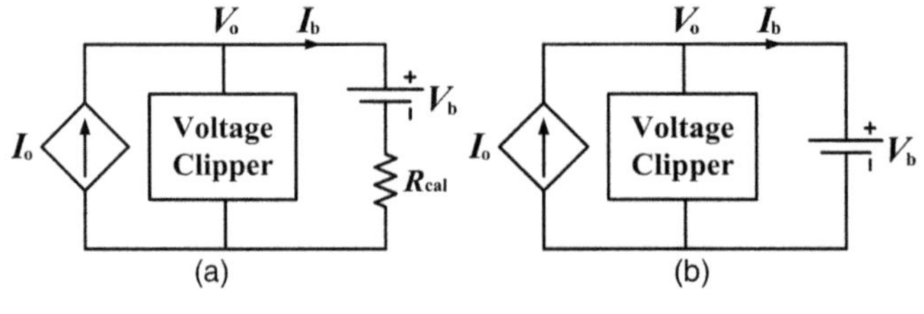

FIGURE 5.10
Calibration circuits during the (a) calibration process and (b) charging process.

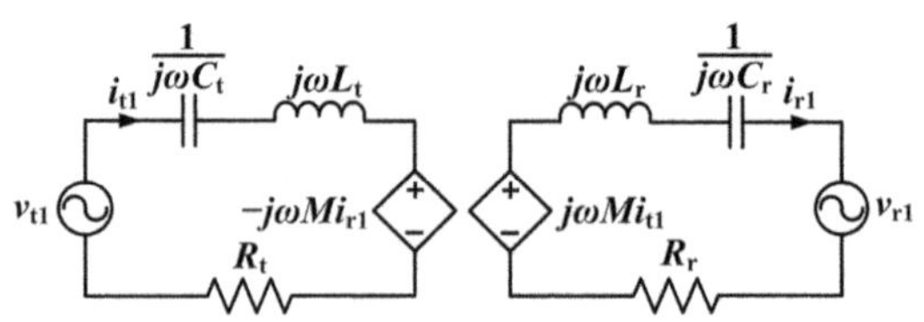

FIGURE 5.11
Equivalent circuit of the SS-compensated resonators.

According to Kirchhoff's voltage law,

$$v_{t1} = Z_t i_{t1} - j\omega M i_{r1} \tag{5.9}$$

$$-v_{r1} = Z_r i_{r1} - j\omega M i_{t1} \tag{5.10}$$

where $Z_t = j\omega L_t + \dfrac{1}{j\omega C_t} + R_t$ and $Z_r = j\omega L_r + \dfrac{1}{j\omega C_r} + R_r$. Then, i_{r1} and v_{r1} can be estimated based on the fundamental components of i_{t1} and v_{t1} as

$$i_{r1} = \frac{Z_t}{j\omega M} i_{t1} - \frac{1}{j\omega M} v_{t1} \tag{5.11}$$

$$v_{r1} = -\frac{\omega^2 M^2 + Z_t Z_r}{j\omega M} i_{t1} + \frac{Z_r}{j\omega M} v_{t1} \tag{5.12}$$

The amplitudes of transmitter components (i.e., I_{t1} and V_{t1}) can be further calculated based on Equations (5.11) and (5.12) as

$$I_{r1} = \left| \frac{Z_t}{j\omega M} i_{t1} - \frac{1}{j\omega M} v_{t1} \right| \tag{5.13}$$

$$V_{r1} = \left| \frac{Z_r}{j\omega M} v_{t1} - \frac{\omega^2 M^2 + Z_t Z_r}{j\omega M} i_{t1} \right| \tag{5.14}$$

where the symbol $|\cdot|$ represents the amplitude of an AC signal. According to Fourier analysis, I_{s1} and V_{s1} are proportional to the output current and voltage (i.e., I_o and V_o) as

$$I_{r1} = \frac{\pi I_o}{2} \tag{5.15}$$

$$V_{r1} = \frac{4(V_o + 2V_D)}{\pi} \tag{5.16}$$

where V_D represents the diode forward voltage. Here, the diode forward voltages are assumed to be identical for all diodes in the rectifier, which does not affect the generality of the analysis. Substituting Equations (5.15) and (5.16) into Equations (5.13) and (5.14), respectively, yields

$$I_o = \frac{2}{\pi} \left| \frac{Z_t}{j\omega M} i_{t1} - \frac{1}{j\omega M} v_{t1} \right| \tag{5.17}$$

$$V_o = \frac{\pi}{4}\left|\frac{Z_r}{j\omega M}v_{t1} - \frac{\omega^2 M^2 + Z_t Z_r}{j\omega M}i_{t1}\right| - 2V_D \tag{5.18}$$

Typically, the resonant frequencies of both resonators are designed to be close to the operating frequency. As a result, the resonator impedances Z_t and Z_r can be approximated by their ESRs, i.e., $Z_t = R_t$ and $Z_r = R_r$. Moreover, for coupled coils with high quality (Q) factors, the ESRs of the resonators are negligible compared to other terms in Equations (5.17) and (5.18). Consequently, they can be simplified as

$$I_o \approx \frac{2}{\pi\omega M}V_{t1} \tag{5.19}$$

$$V_o \approx \frac{\pi\omega M}{4}I_{t1} - 2V_D \tag{5.20}$$

where V_{t1} and I_{t1} represent the amplitudes of v_{t1} and i_{t1}, respectively. During the current charging stage, the voltage clipper remains inactive, and the battery charging current is equal to the output current (i.e., $I_b = I_o$). In the voltage charging stage, the battery charging voltage matches the output voltage (i.e., $V_b = V_o$). Consequently, the charging current and voltage can be estimated on the transmitter side using

$$I_b = \frac{2}{\pi\omega M}V_{t1} \tag{5.21}$$

$$V_b = \frac{\pi\omega M}{4}I_{t1} - 2V_D \tag{5.22}$$

Notably, the mutual inductance M is unknown. However, it is not necessary to determine M using the proposed independent auxiliary circuit-based control scheme. During the calibration stage, gradually decreasing the phase shift angle in fixed small steps incrementally increases v_{t1}. Consequently, I_b also increases. Since S_{cal} is turned off, the output voltage V_o, which is the sum of the battery voltage and the voltage across the calibration resistor (i.e., $V_o = V_b + I_b R_{cal}$), rises until it reaches the upper limit (i.e., $V_o = V_{oH}$). At this point, the corresponding I_{t1} saturates at I_{t1sat}. Here, V_{oH} and V_D are predetermined values, while I_{t1sat} can be directly measured on the transmitter side. Then, the transmitter-side estimation of the output charging current and voltage can be obtained using Equations (5.21) and (5.22) as

$$I_b = \frac{I_{t1sat}}{2(V_{oH} + 2V_D)}V_{t1} \tag{5.23}$$

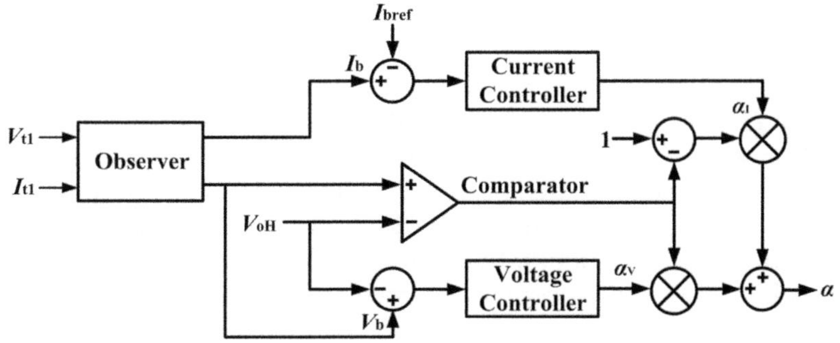

FIGURE 5.12
Control block diagram of a typical transmitter-side controller with automatic current and voltage charging selection.

$$V_b = \frac{(V_{oH} + 2V_D)}{I_{t1sat}} I_{t1} - 2V_D \tag{5.24}$$

Based on the observer-based estimations, a transmitter-side control can be implemented. The control block diagram is shown in Figure 5.12. The digital controller intuitively integrates both current and voltage control. When the estimated battery voltage V_b is below the upper limit V_{oH}, the comparator output is 0, and the phase shift angle of the transmitter-side inverter is regulated by the current controller (i.e., $\alpha = \alpha_I$), ensuring that I_b tracks the reference current I_{bref}. Conversely, when the estimated V_b reaches the upper limit V_{oH}, the comparator output switches to 1, and the phase shift angle is controlled by the voltage controller (i.e., $\alpha = \alpha_V$), maintaining V_b at the upper limit V_{oH}. Both voltage and current control can be implemented using either linear or nonlinear methods. If conventional PI controllers are used, their expressions are given by

$$\alpha_V = K_{pV}(V_b - V_{oH}) + \int K_{iV}(V_b - V_{oH})dt \tag{5.25}$$

$$\alpha_I = K_{pI}(I_b - I_{bref}) + \int K_{iI}(I_b - I_{bref})dt \tag{5.26}$$

where K_{pV} and K_{iV} are the proportional and integral gains of the voltage controller, respectively, and K_{pI} and K_{iI} are the proportional and integral gains of the current controller, respectively.

Although the proposed independent auxiliary circuit-based control scheme eliminates the need for transmitter-side monitoring of receiver-side parameters, the mutual inductance M can still be estimated in real time using the following equation derived from Equation (5.22):

$$M = \frac{4(V_{oH} + 2V_D)}{\pi \omega I_{t1sat}} \tag{5.27}$$

5.2.1.4 Control Algorithm Flowchart

Figure 5.13 presents the flowchart that illustrates the two processes involved in the proposed control implementation for the WPT system utilizing the calibration circuit. The pre-calibration stage serves to transmit the message "calibration will start soon" from the transmitter to the receiver. During the calibration stage, the estimated output charging current and output charging voltage are determined. Additionally, the mutual inductance can be monitored, though this is optional. Following calibration, the system enters the pre-charging stage, which signals the receiver with the message "charging will start soon". Finally, the charging stage commences. During the charging stage, the system operates at the resonant frequency, with the battery's charging current and voltage regulated by the transmitter-side control, as outlined in Figure 5.12. The charging process concludes when the phase shift angle of the proposed control exceeds a predefined threshold (i.e., 175°), ensuring the charging current remains minimal for a fully charged battery. In this

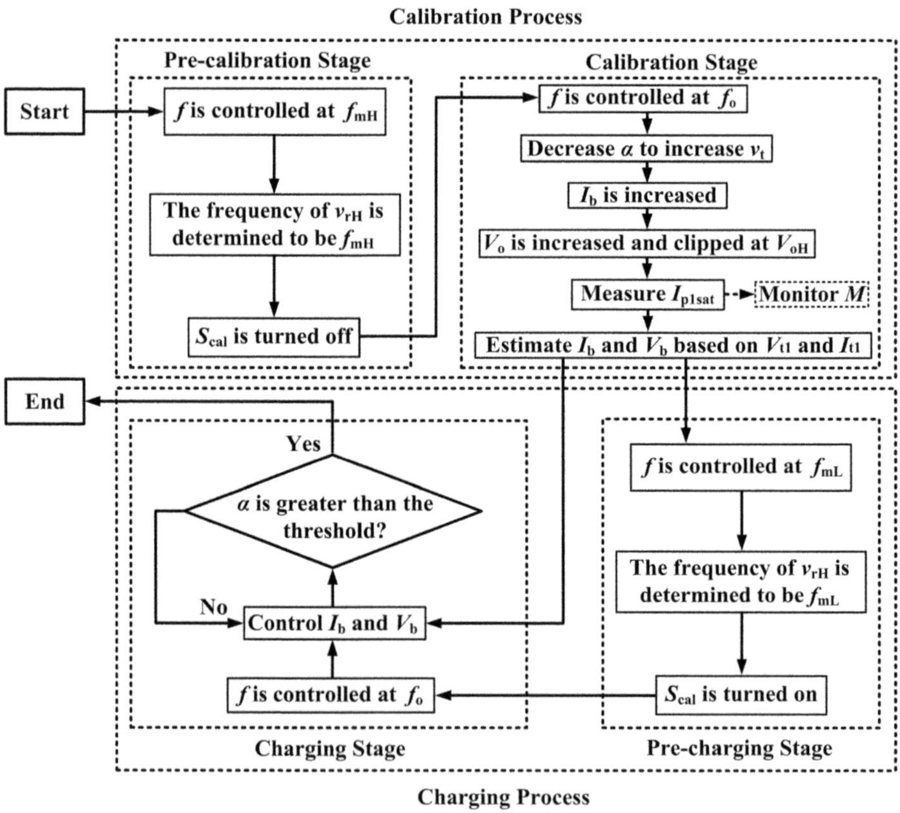

FIGURE 5.13
Flowchart of the control algorithm for the WPT system utilizing the calibration circuit.

algorithm, I_{t1} can also be used as an indicator for misalignment detection, as it is already utilized for both calibration and charging. Specifically, a misalignment event causes a sudden shift in I_{t1}. If the steady-state deviation from its nominal value exceeds a predefined threshold, the system identifies the coupled coils as misaligned. The threshold is empirically determined based on system parameters and application requirements, with a typical value to ensure reliable misalignment detection while minimizing false triggers. To prevent error propagation, misalignment detection can be delayed for a brief period to allow I_{t1} to stabilize. Once misalignment is confirmed, both pre-calibration and calibration are performed to estimate the new output charging current and charging voltage.

5.2.1.5 Simulation Results

The simulations are performed using PSIM 10.0. The key specifications of the WPT system are summarized in Table 5.3. The switching frequency is adjusted within the range of 90–110 kHz to comply with the "Qi" standard (85–205 kHz). During the calibration stage, the phase shift angle is decreased

TABLE 5.3

Main Specifications of the WPT System with the Auxiliary Circuit

Parameter	Value
Transmitter and receiver inductances L_t, L_r	48.4 μH
Transmitter capacitance C_t	52.5 nF
Receiver capacitance C_r	52.3 nF
Transmitter and receiver resistance R_t, R_r	0.05 Ω
Nominal mutual inductance of the coupled coils M	7.1 μH
Input DC voltage V_{DC}	12 V
Diode forward voltage V_D	0.5 V
Capacitance of the filter capacitor C_o	100 μF
Resistance for the calibration R_{cal}	0.5 Ω
Upper output voltage limit of the rectifier V_{oH}	4.2 V
Initial battery voltage V_b	3.6 V
Nominal resonant frequency f_o	100 kHz
High reference frequency f_{mH}	110 kHz
Low reference frequency f_{mL}	90 kHz
Initial α during the calibration α_{cal0}	150°
Steps of α variations during the calibration $\Delta\alpha_{cal}$	–30°
Frequency of α change during the calibration f_{cal}	500 Hz
Charging current reference I_{bref}	1.67 A
Charging voltage reference V_{bref}	4.2 V
Voltage divider resistance R_{rH1}	9 kΩ
Voltage divider resistance R_{rH2}	1 kΩ
Tolerance of transmitter current peak variations	5%

from the initial 150° to 0° by a fixed step of –30° at every 2 ms. The simulation implementation of the calibration circuit is depicted in Figure 5.14. The frequency of the nodal voltage v_{rH} (i.e., f_m) is determined as half of the reciprocal of the pulse-width counter output. Both voltage and current controllers utilize conventional PI controllers, with proportional and integral gains set to 10 and 10,000, respectively (i.e., $K_{pV} = 10$, $K_{iV} = 10{,}000$, $K_{pI} = 10$, and $K_{iI} = 10{,}000$).

Figure 5.15 shows the waveforms of key parameters during the pre-calibration, calibration, pre-charging, and charging stages. These include the switching frequency, the frequency of the nodal voltage v_{sH} (i.e., f_m), the phase shift angle (i.e., α), the output of the frequency detector (i.e., S_{cal}), and the fundamental component of the transmitter current (i.e., i_{t1}) along with its peak value (i.e., I_{t1}). During the pre-calibration phase (0–0.001 s), the switching frequency is maintained at $f_{mH} = 110$ kHz, resulting in a measured nodal voltage frequency of 110 kHz. Consequently, the frequency detector outputs 0, keeping the switch S_{cal} off. At 0.001 s, the calibration phase begins, and the switching frequency gradually transitions from 110 kHz to the

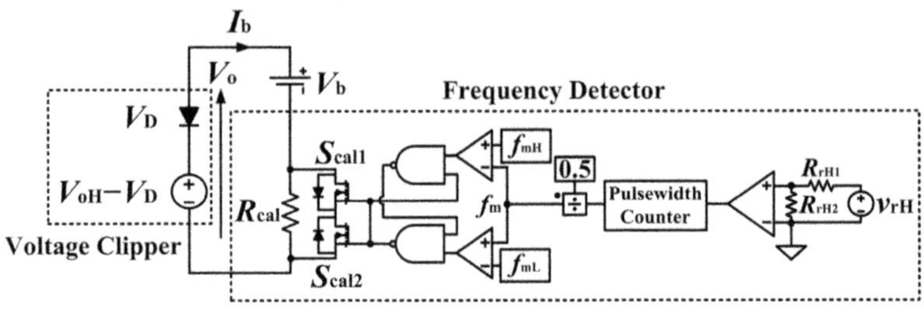

FIGURE 5.14
Schematic diagram of the calibration circuit in simulation.

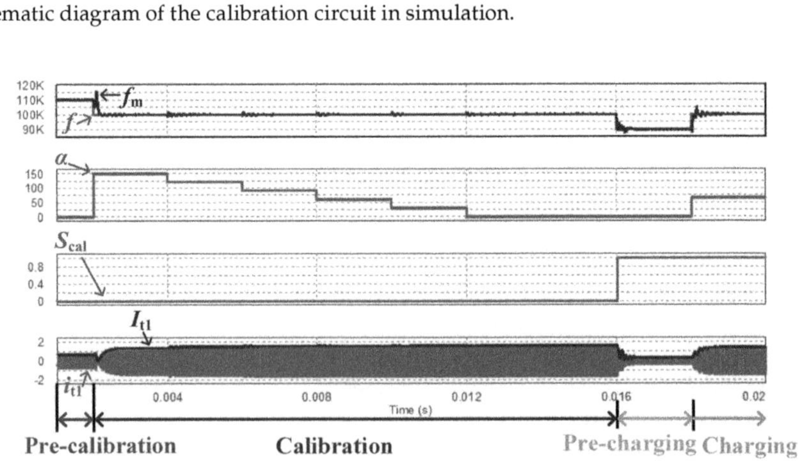

FIGURE 5.15
Simulated waveforms of the WPT system with the calibration circuit during the pre-calibration, calibration, and pre-charging stages. (Adapted from reference [2].)

resonant frequency 100 kHz, while the output of the frequency detector remains unchanged, keeping S_{cal} off. From 0.001 to 0.016 s, the phase shift angle decreases incrementally from 150° to 0° in fixed steps of –30°, causing the peak transmitter current (I_{t1}) to rise until saturation. The saturated peak current (I_{t1sat}) is observed at 1.5 A. By applying Equations (5.23) and (5.24) along with the parameters in Table 5.3, the output charging current and voltage can be estimated as $I_b = 0.144V_{t1}$ and $V_b = 3.467I_{t1} - 1$, where V_{t1} and I_{t1} represent the amplitudes of the fundamental components measured at the transmitter side. Notably, the mutual inductance can also be estimated at this stage using Equation (5.27) and the parameters in Table 5.3. The estimated mutual inductance is 7.02 μH, with a relative error of only 1.13% compared to the actual value of 7.1 μH. The complete calibration process requires only 16 ms to determine the mutual inductance. At 0.016 s, the pre-charging phase begins, lasting until 0.018 s. During this phase, the switching frequency is set to $f_{mL} = 90$ kHz, leading to a detected nodal voltage frequency of 90 kHz. As a result, the frequency detector outputs 1, turning on S_{cal}. The charging phase starts at 0.018 s, during which the switching frequency shifts from 90 kHz to the resonant frequency $f_o = 100$ kHz. The frequency detector output remains unchanged, ensuring that the calibration resistor R_{cal} is bypassed.

Figure 5.16 exhibits the one-cycle charging profile of the system, showing the charging current and voltage waveforms under the proposed control strategy. In the simulation, the practical one-hour duration is scaled to 0.005 s, meaning that the full battery charge, starting from the nominal condition (i.e., $V_b = 3.6$ V), takes approximately 2.4 hours in practice (equivalent to 0.012 s in simulation). The charging profile demonstrates that both the

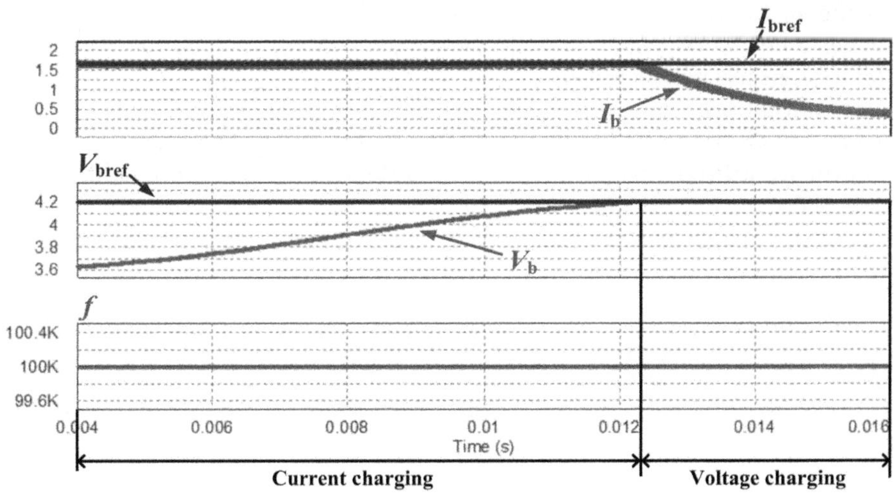

FIGURE 5.16
One-cycle charging profile of the WPT system with the auxiliary circuit in simulation. (Adapted from reference [2].)

charging current and voltage are precisely regulated to follow their respective references (i.e., $I_{bref}=1.67$ A and $V_{bref}=4.2$ V). The transition from the current charging mode to the voltage charging mode occurs smoothly when the battery voltage reaches the threshold. Additionally, the switching frequency remains at the resonant frequency throughout the entire charging process.

The simulated waveforms of v_t and i_t at approximately 0.008 s during the current charging period and 0.014 s during the voltage charging period are shown in Figure 5.17. The phase shift angle is regulated to 77.72° at 0.008 s and 104.77° at 0.014 s, respectively. Additionally, Figure 5.17a presents the corresponding battery charging current along with the estimated charging current based on the transmitter-side observer (i.e., I_{best}) at around 0.008 s, with a relative estimation error of less than 1%. Similarly, Figure 5.17b shows the corresponding battery charging voltage and the estimated charging voltage (i.e., V_{best}) at approximately 0.014 s, also with a relative estimation error of less than 1%.

Simulations are also performed on the WPT system under misalignment conditions. Figure 5.18 presents the waveforms of the operating frequency (i.e., f), the frequency of the nodal voltage (i.e., f_m), the phase shift angle (i.e., α), the switching signal for the calibration resistor (i.e., S_{cal}), the fundamental transmitter current (i.e., i_{t1}), and its peak value (i.e., I_{t1}) when the mutual inductance decreases from the nominal 7.1 to 4.8 μH, corresponding to a coupling coefficient shift from 0.147 to 0.1. Similarly, Figure 5.19 exhibits the corresponding waveforms for the system when the mutual inductance increases from 7.1 to 9.02 μH, resulting in a coupling coefficient shift from 0.147 to 0.186. From 0.015 to 0.02 s, both systems operate at 100 kHz with $\alpha=77.72°$ under nominal mutual inductance (i.e., $M=7.1$ μH). At 0.02 s, misalignment occurs, causing I_{t1} to change. A 0.002 s misalignment detection period ensures I_{t1} reaches a new steady state. In both cases, I_{t1} variations exceed the 5% preset tolerance, necessitating recalibration.

Pre-calibration runs from 0.022 to 0.024 s, adjusting the operating frequency f from 100 to 110 kHz with $\alpha=0°$. The frequency of the nodal voltage f_m shifts accordingly, triggering S_{cal} to toggle and initiating pre-calibration.

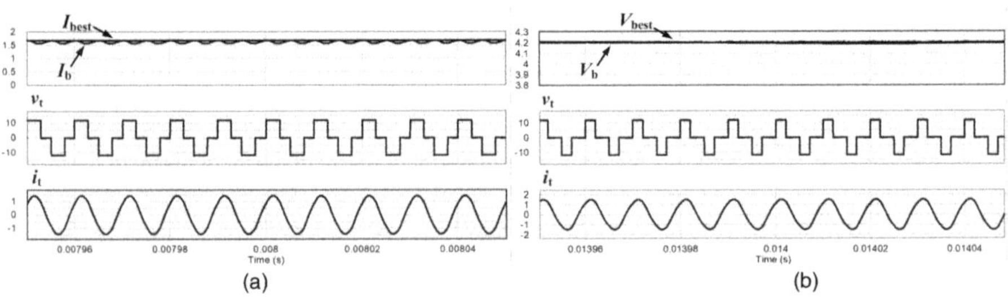

FIGURE 5.17
Simulated waveforms of key parameters in the WPT system with the auxiliary circuit during (a) the current charging stage at 0.008 s and (b) the voltage charging stage at 0.014 s. (Adapted from reference [2].)

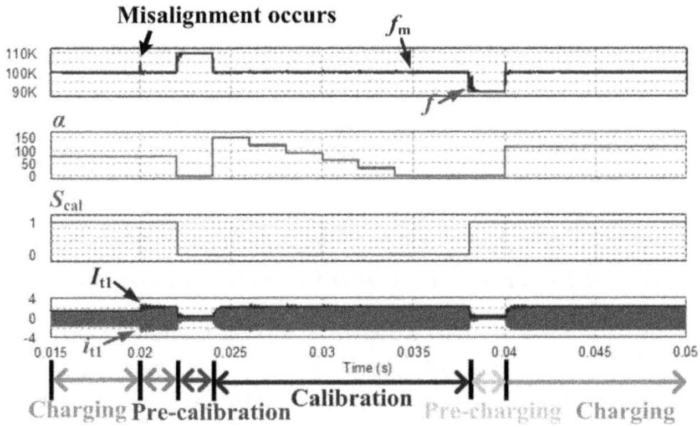

FIGURE 5.18
Simulated waveforms of key parameters in the WPT system with the auxiliary circuit when the mutual inductance decreases from the nominal 7.1 to 4.8 μH. (Adapted from reference [2].)

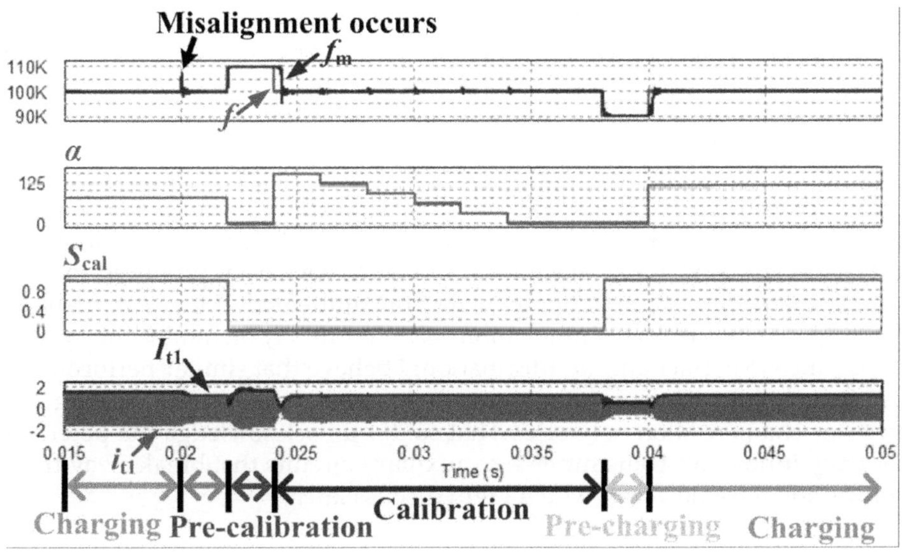

FIGURE 5.19
Simulated waveforms of key parameters in the WPT system with the auxiliary circuit when the mutual inductance increases from the nominal 7.1 to 9.02 μH. (Adapted from reference [2].)

Calibration begins at 0.024 s, maintaining $f = 100$ kHz until 0.038 s. Here, α decreases stepwise from 150° to 0° in –30° increments, while I_{t1} rises until saturation. The saturated peak currents (i.e., I_{t1sat}) are 2.22 and 1.18 A, as shown in Figures 5.18 and 5.19, respectively. This results in the estimated output charging current and voltage being $I_b = 0.231V_{t1}$ and $V_b = 2.342I_{t1} - 1$ for $I_{t1sat} = 2.22$ A, and $I_b = 0.113V_{t1}$ and $V_b = 4.407I_{t1} - 1$ for $I_{t1sat} = 1.18$ A. Meanwhile, the mutual inductances can be estimated to be 4.74 and 8.93 μH, with relative errors of only 1.25% and 1% compared to the actual values. The entire calibration process is completed within 16 ms. After calibration, pre-charging starts at 0.038 s with $f = 90$ kHz and $\alpha = 0°$, toggling S_{cal} from 0 to 1 over 0.002 s. The charging stage then commences using the updated observer equations, with R_{cal} bypassed.

5.2.1.6 Evaluation of the Auxiliary Circuit-Based Control Scheme

Experimental validations of the auxiliary circuit-based control scheme are provided in [2], demonstrating superior performance compared to the dependent transmitter-side control in [9], particularly for coupled resonators with high Q factors. The main advantage of the auxiliary circuit-based control scheme lies in its ability to dynamically calibrate the system status across a wide range of operating conditions. By periodically updating the system status, the transmitter-side control achieves high accuracy, effectively adapting to real-time variations in load, mutual coupling misalignments, parameter drifts, and unknown disturbances. In some aspects, the auxiliary circuits in the auxiliary circuit-based control scheme function as additional modules to account for uncertainties, enabling more precise control at the transmitter side. However, a simpler circuit design is generally preferred. The most cost-effective alternative is to employ purely software-based approaches, such as AI, to achieve the same objective. Nevertheless, accuracy remains the primary trade-off. While the proposed auxiliary circuit limits the number of additional components on the receiver side, their presence cannot be entirely eliminated. Some alternative strategies have been proposed to shift the receiver-side auxiliary circuits to the transmitter side. However, many of these approaches lack true innovation, as they still rely on modifying system impedance to extract output information. I believe that similar performance can be achieved through improvements in control algorithms, driving circuits, or compensation networks. That said, I do acknowledge the potential for truly innovative transmitter-side auxiliary circuits that break away from conventional frameworks, which I will discuss later.

5.2.2 Self-Oscillating Control Scheme

5.2.2.1 Background of Self-Oscillating Control Strategies

Self-oscillating control in WPT systems is a transmitter-side control method that synchronizes the inverter driving signals with the receiver current phase angles without requiring modulation or complex computations [10]. For next-generation WPT systems in high-frequency and high-power applications, self-oscillating control provides distinct advantages over conventional PI and MPC. These include simplified hardware implementation, enhanced robustness against parameter drifts, easier tuning, and greater cost-effectiveness [11].

Extensive studies have explored conventional self-oscillating control for various converters to achieve load-independent outputs in two-stage WPT systems [12–16]. The self-oscillating control for achieving load-independent outputs in IPT systems is introduced in [12], where a phase detection contactless current transformer is utilized to detect the phase angle of the secondary current, serving as the reference for the transmitter-side inverter's driving signal. Building on this concept, [10] proposes an integrated transformer that combines power transfer and current sensing to enhance output controllability and system adaptability under misalignment conditions. In [13], an enhanced self-oscillating control method is developed, incorporating an integrated four-coil transformer and a PI controller to enable precise output voltage regulation with rapid and accurate adjustments. The work in [14] applies self-oscillating control to two-stage series-parallel (SP) and parallel-series (PS) compensated WPT systems to achieve constant voltage (CV) outputs. Furthermore, self-oscillating control is extended to LCC-LCC compensated WPT systems in [15] to realize quasi-load-independent outputs. A comprehensive investigation of self-oscillating control is presented in [16], covering Litz wire-based low-frequency WPT systems with SS, SP, PS, parallel-parallel (PP), LCC-S, and LCC-LCC compensation networks, as well as copper-based high-frequency WPT systems with SS and SP compensation networks. In [17], self-oscillating control is applied to domino WPT systems to achieve quasi-load-independent outputs.

5.2.2.2 Analysis of General WPT Systems with Self-Oscillating Control

The schematic diagram of a typical two-stage WPT system includes a DC power supply, a full-bridge DC/AC inverter, coupled resonators with coils and compensation networks, a diode-bridge rectifier, a filtering capacitor, and a resistive load emulating a battery, as shown in Figure 5.20. Here, v_t and v_r represent the input and output voltages of the resonators, while i_t and

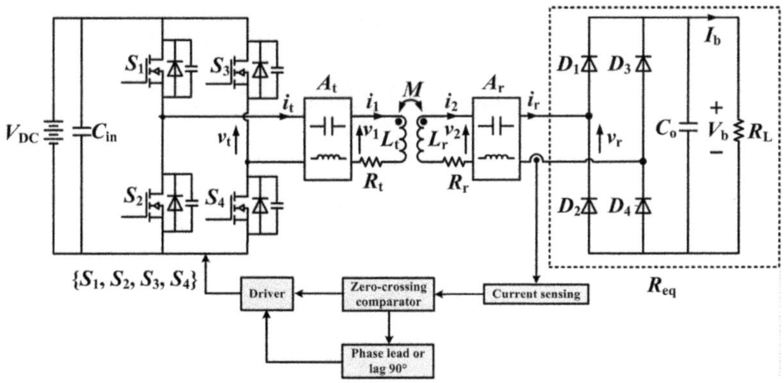

FIGURE 5.20
Schematic diagram of a typical two-stage WPT system with self-oscillating control.

i_r denote the transmitter and receiver currents. L_t and L_r are the self-inductances of the transmitter and receiver coils, respectively, with M as the mutual inductance. R_t and R_r correspond to the ESRs of the resonators. R_{eq} is the equivalent load resistance, which accounts for the diode-bridge rectifier, filtering capacitor, and resistive load. Compensation networks are employed to offset the leakage inductances of the coupled coils. In SS compensation topology, the transmitter-side and receiver-side compensation networks consist of capacitors C_t and C_r, respectively, both connected in series. The capacitance values are designed to satisfy the resonance condition $\omega_o = \frac{1}{\sqrt{L_t C_t}} = \frac{1}{\sqrt{L_r C_r}}$, where ω_o is the resonant angular frequency.

The general matrix-based representations of the compensation networks (i.e., A_t and A_r) are provided as

$$\begin{bmatrix} v_t \\ i_t \end{bmatrix} = A_t \begin{bmatrix} v_1 \\ i_1 \end{bmatrix} = \begin{bmatrix} a_{t11} & a_{t12} \\ a_{t21} & a_{t22} \end{bmatrix} \begin{bmatrix} v_1 \\ i_1 \end{bmatrix} \tag{5.28}$$

$$\begin{bmatrix} v_2 \\ i_2 \end{bmatrix} = A_r \begin{bmatrix} v_r \\ i_r \end{bmatrix} = \begin{bmatrix} a_{r11} & a_{r12} \\ a_{r21} & a_{r22} \end{bmatrix} \begin{bmatrix} v_r \\ i_r \end{bmatrix} \tag{5.29}$$

Here, v_1 and i_1 represent the output voltage and current of the transmitter-side compensation network, while v_2 and i_2 denote the input voltage and current of the receiver-side compensation network. By combining Equations (5.28) and (5.29), v_t can be expressed as

$$v_t = \left(\alpha R_{eq} + \beta\right) i_r \tag{5.30}$$

where

$$\alpha = \frac{\gamma a_{r11} + \delta a_{r21}}{j\omega M} \tag{5.31}$$

$$\beta = \frac{\gamma a_{r12} + \delta a_{r22}}{j\omega M} \tag{5.32}$$

$$\gamma = a_{t11} Z_t + a_{t12} \tag{5.33}$$

$$\delta = \gamma Z_r + a_{t11} \omega^2 M^2 \tag{5.34}$$

$$Z_t = j\omega L_t + R_t \tag{5.35}$$

$$Z_r = j\omega L_r + R_r \tag{5.36}$$

Based on Equation (5.31), α is regulated to be zero for the WPT system in achieving constant current (CC) output as

$$\frac{i_r}{v_t} = \frac{1}{\beta} \tag{5.37}$$

where β is adjusted to zero in the WPT system to achieve a CV output as

$$\frac{v_r}{v_t} = \frac{1}{\alpha} \tag{5.38}$$

The properties (i.e., real, imaginary, or zero) of major components in Equations (5.31)–(5.36) are summarized in Table 5.4. Here, α is a real number, while β is an imaginary number. Consequently, CC output in WPT systems can be achieved by controlling the phase angle of v_t to maintain a 90° phase difference with i_r. Similarly, CV output can be realized by ensuring v_t is either in phase or in reverse phase with i_r. The conventional self-oscillating control of WPT systems is designed based on this fundamental principle.

The outputs of WPT systems with the four basic compensation networks (i.e., SS, SP, PS, and PP) remain load-independent under CC and CV modes when the ESRs of the coupled resonators are neglected. The ideal operating frequencies for achieving CC and CV outputs in the SS-compensated networks can be determined as

TABLE 5.4

Properties of Major Components in Equations (5.31)–(5.36)

Component	Property
a_{r11}	Real
a_{r12}	Imaginary or 0
a_{r21}	Imaginary or 0
a_{r22}	Real
a_{t11}	Real
a_{t12}	Imaginary or 0
a_{t21}	Imaginary or 0
a_{t22}	Real
γ	Imaginary
δ	Real
α	Real
β	Imaginary

$$\omega_{\mathrm{CC\_SS}} = \omega_{\mathrm{o}} \tag{5.39}$$

$$\omega_{\mathrm{CVH\_SS}} = \omega_{\mathrm{o}} \sqrt{\frac{\sqrt{L_{\mathrm{t}} L_{\mathrm{r}}}}{\sqrt{L_{\mathrm{t}} L_{\mathrm{r}}} - M}} \tag{5.40}$$

$$\omega_{\mathrm{CVL\_SS}} = \omega_{\mathrm{o}} \sqrt{\frac{\sqrt{L_{\mathrm{t}} L_{\mathrm{r}}}}{\sqrt{L_{\mathrm{t}} L_{\mathrm{r}}} + M}} \tag{5.41}$$

where $\omega_{\mathrm{CVH\_SS}}$ and $\omega_{\mathrm{CVL\_SS}}$ denote the operating angular frequencies of the SS-compensated WPT system required to achieve CV outputs, with $\omega_{\mathrm{CVH\_SS}}$ being above the resonant angular frequency and $\omega_{\mathrm{CVL\_SS}}$ being below it. The corresponding ideal output current in the CC mode and ideal output voltages in the CV mode can be expressed as

$$i_{\mathrm{r}} = j \frac{v_{\mathrm{t}}}{\omega_{\mathrm{CC\_SS}} M} \tag{5.42}$$

$$v_{\mathrm{rH}} = \sqrt{\frac{L_{\mathrm{r}}}{L_{\mathrm{t}}}} v_{\mathrm{t}} \tag{5.43}$$

$$v_{\mathrm{rL}} = -\sqrt{\frac{L_{\mathrm{r}}}{L_{\mathrm{t}}}} v_{\mathrm{t}} \tag{5.44}$$

From Equation (5.42), it can be observed that a load-independent CC output in an SS-compensated WPT system can be achieved by ensuring the phase angle of v_{t} lags 90° behind the measured phase angle of i_{t}. Additionally, the amplitude of v_{r} is determined by the amplitude of i_{r} multiplied by $\omega_{\mathrm{CC\_SS}}$ and M, while the system's operating frequency must satisfy Equation (5.39). Similarly, to achieve a load-independent CV output in an SS-compensated WPT system, the phase differences between v_{r} and v_{t} must satisfy Equation (5.43) or (5.44), while the system's operating frequency must comply with the corresponding Equation (5.40) or (5.41). The phase angle of v_{t} also corresponds to the phase angle of the switching signal S_1. Therefore, the output square waveforms from the comparator or the 90° compensation circuit in Figure 5.20 can be directly fed into the driver to control the transmitter-side inverter. Regarding the amplitude relationships between i_{r} and v_{t}, as well as v_{r} and v_{t}, these can be further utilized to establish the relationships between I_{b} and v_{t}, and V_{b} and v_{t}, based on Equations (5.15) and (5.16), respectively. Figure 5.21 presents the simulation results of I_{b} and V_{b} for an SS-compensated WPT system with key parameters: $L_{\mathrm{t}} = L_{\mathrm{r}} = 48.2$ μH, $C_{\mathrm{t}} = C_{\mathrm{r}} = 52.5$ nF, $M = 7.1$ μH, and $V_{\mathrm{dc}} = 11.1$ V under various load conditions. As observed, the load-independent

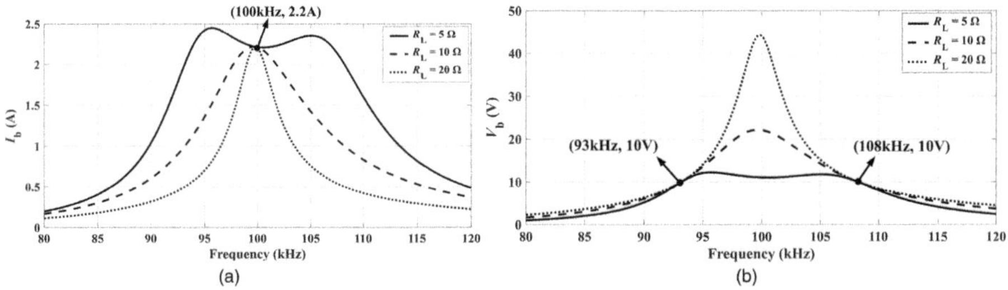

FIGURE 5.21
Simulated output of (a) I_b and (b) V_b for an SS-compensated WPT system operating under load-independent conditions across various load scenarios.

operating point for CC output occurs precisely at the resonant frequency (according to Equation (5.39)), whereas the load-independent operating points for CV output appear on either side of the resonant frequency (according to Equations (5.40) and (5.41)). The corresponding amplitude values of I_b and V_b also confirm the validity of the analysis in Equations (5.42)–(5.44).

The ideal operating frequencies for achieving CC and CV outputs in SP-compensated networks can be determined as

$$\omega_{CCH\_SP} = \omega_o \sqrt{\frac{\sqrt{L_t L_r}}{\sqrt{L_t L_r} - M}} \tag{5.45}$$

$$\omega_{CCL\_SP} = \omega_o \sqrt{\frac{\sqrt{L_t L_r}}{\sqrt{L_t L_r} + M}} \tag{5.46}$$

$$\omega_{CV\_SP} = \omega_o \sqrt{\frac{L_t L_r}{L_t L_r - M^2}} \tag{5.47}$$

The corresponding ideal output currents in the CC mode and ideal output voltage in the CV mode can be obtained as

$$i_{rH} = j \frac{v_t}{\omega_{CCH\_SP}\left(M - \sqrt{L_t L_r}\right)} \tag{5.48}$$

$$i_{rL} = j \frac{v_t}{\omega_{CCL\_SP}\left(M + \sqrt{L_t L_r}\right)} \tag{5.49}$$

$$v_r = \frac{L_r}{M} v_t \tag{5.50}$$

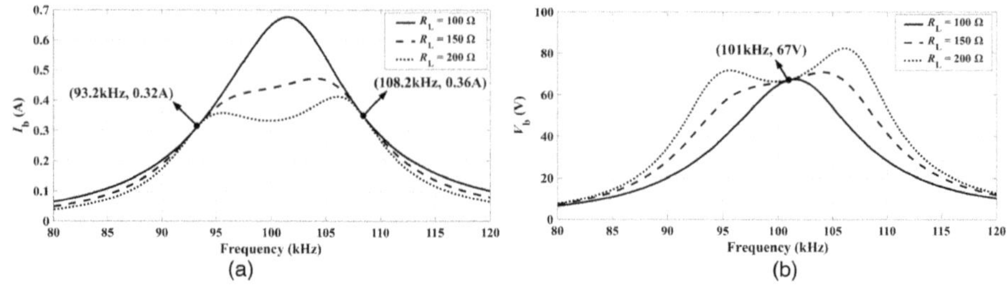

FIGURE 5.22
Simulated output of (a) I_b and (b) V_b for an SP-compensated WPT system operating under load-independent conditions across various load scenarios.

Figure 5.22 displays the simulated results of I_b and V_b for an SP-compensated WPT system with the following key parameters: $L_t = L_r = 48.2$ μH, $C_t = C_r = 52.5$ nF, $M = 7.1$ μH, and $V_{dc} = 11.1$ V under varying load conditions. The results demonstrate that the load-independent CV output occurs at the frequency that can be calculated based on Equation (5.47), while the load-independent CC outputs manifest on either side of the resonant frequency, which can be calculated based on Equations (5.45) and (5.46). The corresponding amplitudes of I_b and V_b corroborate the theoretical analysis in Equations (5.48)–(5.50).

For PS-compensated networks, there is no single optimal operating frequency to achieve CC output, whereas a specific resonant frequency exists for attaining CV output as

$$\omega_{CV\_PS} = \omega_o \sqrt{\frac{L_t L_r}{L_t L_r - M^2}} \tag{5.51}$$

The corresponding ideal output voltage in the CV mode can be derived as

$$v_r = \frac{M}{L_t} v_t \tag{5.52}$$

Figure 5.23 presents the simulated V_b for a PS-compensated WPT system with main parameters: $L_t = L_r = 48.2$ μH, $C_t = C_r = 52.5$ nF, $M = 7.1$ μH, and $V_{dc} = 11.1$ V under varying load conditions. The results confirm that the load-independent CV output occurs at the frequency that can be calculated based on Equation (5.51), with the corresponding V_b values aligning with theoretical analysis in Equation (5.52).

For PP-compensated networks, there is no single optimal operating frequency to achieve CV output, whereas a specific resonant frequency exists for attaining CC output as

$$\omega_{CC\_PP} = \omega_o \sqrt{\frac{L_t L_r}{L_t L_r - M^2}} \tag{5.53}$$

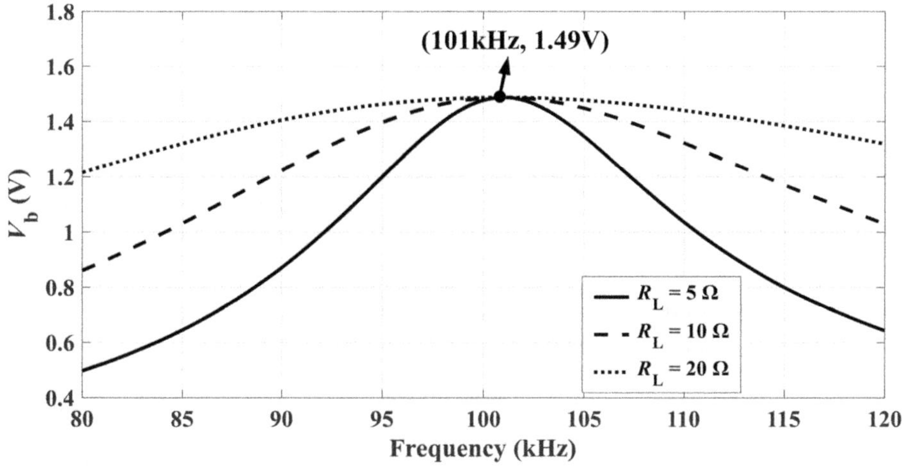

FIGURE 5.23
Simulated V_b for a PS-compensated WPT system operating under the load-independent condition across various load scenarios.

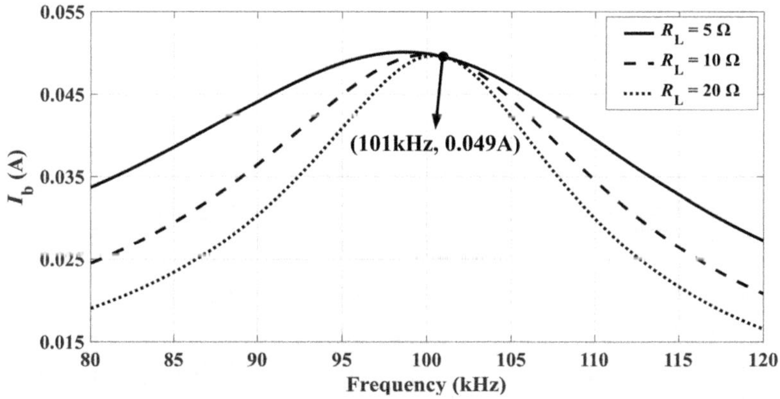

FIGURE 5.24
Simulated I_b for a PP-compensated WPT system operating under the load-independent condition across various load scenarios.

The corresponding ideal output voltage in the CC mode can be derived as

$$i_r = j\frac{v_t}{\omega_{CC\_PP}\left(M - \frac{L_t L_r}{M}\right)} \tag{5.54}$$

Figure 5.24 presents the simulated I_b for a PP-compensated WPT system with main parameters: $L_t = L_r = 48.2$ μH, $C_t = C_r = 52.5$ nF, $M = 7.1$ μH, and $V_{dc} = 11.1$ V under varying load conditions. The results confirm that the load-independent

CC output occurs at the frequency that can be calculated based on Equation (5.53), with the corresponding I_b values aligning with theoretical analysis in Equation (5.54).

The simulation results of the four basic compensation networks in WPT systems demonstrate that load-independent outputs can be achieved by regulating the operating frequencies at specific values. This capability is particularly advantageous for battery charging applications, where CC and CV charging profiles are typically required.

5.2.2.3 Implementation of Communication-Free Self-Oscillating Control Using Pick-up Coil

The self-oscillating control implementation, based on these operating principles, is relatively simple, offering a robust framework for transmitter-side control in next-generation WPT systems with minimal receiver-side complexity. However, a key challenge lies in acquiring feedback information for the receiver current i_r and voltage v_r. While the feedback of v_r can often be omitted, the feedback of i_r remains essential. To address this issue, a pick-up coil, magnetically coupled to the receiver coil but decoupled from the transmitter coil, can be employed.

Figure 5.25 shows the schematic diagram of a typical WPT system, incorporating the pick-up coil to enable self-oscillating control. The current in the

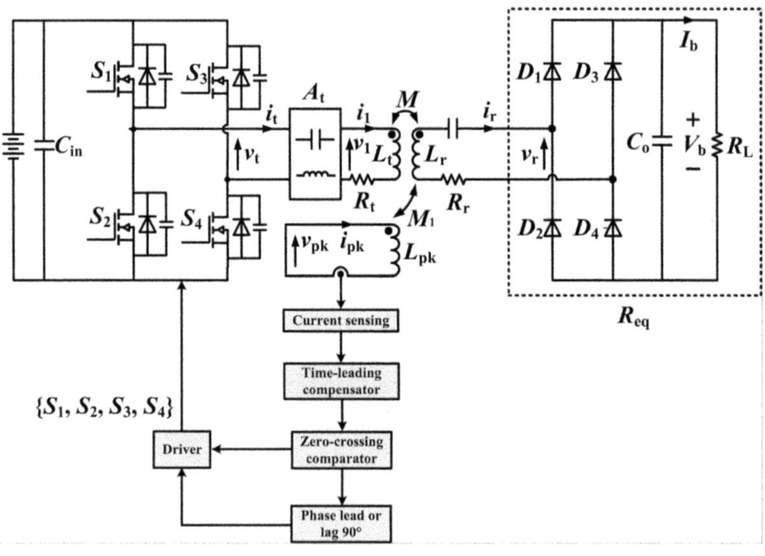

FIGURE 5.25
Schematic diagram of a typical two-stage WPT system with communication-free self-oscillating control using a pick-up coil.

pick-up coil (i.e., i_{pk}) replicates the receiver current (i.e., i_r), allowing the phase information of i_r and v_r to be obtained from i_{pk}. The relationship between i_{pk} and i_r is

$$i_{pk} = \frac{M_1}{L_{pk}} i_r \tag{5.55}$$

where M_1 is the mutual inductance between the receiving coil and the pick-up coil. L_{pk} is the self-inductance of the pick-up coil. However, it is worth noting that the compensation network of the receiving resonator for this method must be series compensation, since i_{pk} can only reflect the current flowing through the receiving coil rather than i_r. Only for the series compensation, the two currents are equalized at the receiver side.

Based on Equation (5.55), the relationship between the amplitude of i_{pk} (i.e., $|i_{pk}|$) and the output current I_b in an SS-compensated WPT system operating in CC mode can be expressed as

$$I_b = \frac{2L_{pk}}{\pi M_1} |i_{pk}| \tag{5.56}$$

Similarly, based on Equations (5.43) or (5.44), the relationship between the amplitudes of v_t (i.e., $|v_t|$) and the output voltage V_b in an SS-compensated WPT system operating in CV mode can be expressed as

$$V_b = \frac{\pi}{4} \sqrt{\frac{L_r}{L_t}} |v_t| - 2V_D \tag{5.57}$$

Based on Equations (5.56) and (5.57), the transmitter-side control references for an SS-compensated WPT system with the communication-free self-oscillating control can be determined as

$$|i_{pkref}| = \frac{\pi M_1}{2L_{pk}} I_{bref} \tag{5.58}$$

$$|v_{tref}| = \frac{4}{\pi} \sqrt{\frac{L_t}{L_r}} (V_{bref} + 2V_D) \tag{5.59}$$

The control block diagram is shown in Figure 5.26. $|i_{pk}|$ and $|v_t|$ are measured and fed back to track the transmitter-side reference values in CC mode and CV mode, respectively. The deviations from these references are processed by the PI controller, which generates the phase shift angle α between the switching signals. Meanwhile, the switching frequency of the transmitter-side inverter is regulated at ω_{CC_SS} in CC mode and at either ω_{CVH_SS} or ω_{CVL_SS} in CV mode to ensure load-independent outputs. Typically,

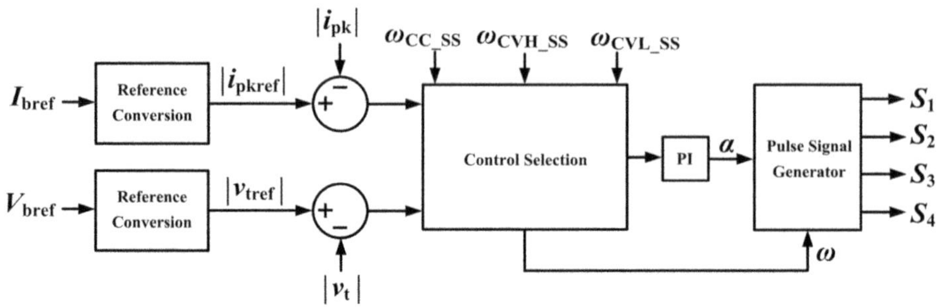

FIGURE 5.26
Control block diagram of communication-free self-oscillating control utilizing a pick-up coil for SS-compensated WPT systems.

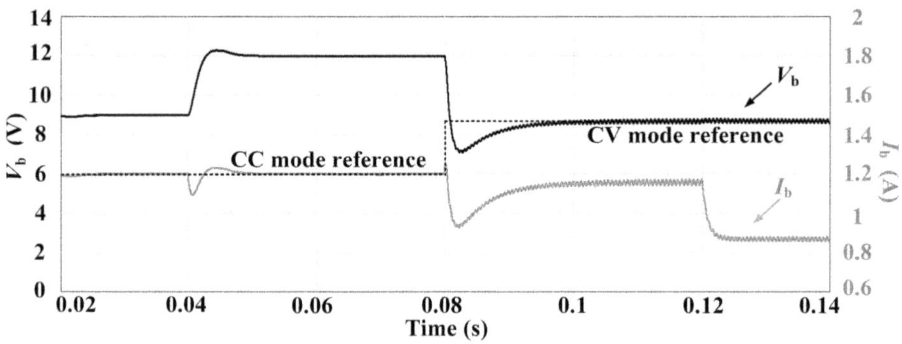

FIGURE 5.27
Simulated waveforms of V_b and I_b for the SS-compensated WPT system with the communication-free self-oscillating control.

ω_{CVH_SS} is preferred over ω_{CVL_SS} as it facilitates easier implementation of soft switching for the transmitter-side inverter.

Figure 5.27 presents the simulated I_b and V_b for an SS-compensated WPT system with a pick-up coil, utilizing the control scheme shown in Figure 5.26. The key system parameters are $L_t = L_r = 48.2$ μH, $C_t = C_r = 52.5$ nF, $M = 7.1$ μH, and $V_{dc} = 10$ V. The diode forward voltage is set to zero, which does not impact the validity of the analysis. The load resistance R_L toggles between 7.5 and 10 Ω at 0.04, 0.08, and 0.12 s. The reference current I_{bref} for CC mode is set to 1.2 A, while the reference voltage V_{bref} for CV mode is set to 8.7 V. The transition from CC mode to CV mode occurs at 0.08 s. The proportional and integral gains of the PI controller for CC mode regulation are $K_p = 30$ and $K_i = 40{,}000$, respectively, while for CV mode regulation, they are $K_p = 30$ and $K_i = 10{,}000$. The results demonstrate that the output current and voltage are precisely regulated to track their respective reference values, even in the presence of load variations. These findings confirm the effectiveness of the communication-free self-oscillating control in SS-compensated WPT systems.

Similar communication-free self-oscillating control strategies can also be applied to PS-compensated WPT systems, with modifications limited to reference conversions and load-independent operating frequencies. However, it is important to note that the voltage-source H-bridge inverter shown in Figure 5.25 cannot be directly used for parallel compensation at the transmitter side. Instead, a current-source inverter is needed. To form a current-source inverter based on the voltage-source H-bridge inverter, an inductor needs to be included at the output of the voltage-source H-bridge inverter. Thus, the PS compensation is essentially an LCL-S compensation. The ideal operating frequencies for achieving CC and CV outputs in LCL-S-compensated networks can be determined as

$$\omega_{\mathrm{CC\_LCLS}} = \sqrt{2}\omega_{\mathrm{o}} \tag{5.60}$$

$$\omega_{\mathrm{CV\_LCLS}} = \omega_{\mathrm{o}} \tag{5.61}$$

The ideal output current in CC mode and the ideal output voltage in CV mode can be derived as

$$i_{\mathrm{r}} = -j\frac{v_{\mathrm{t}}}{\omega_{\mathrm{CC\_LCLS}}M} \tag{5.62}$$

$$v_{\mathrm{r}} = \frac{M}{\sqrt{L_{\mathrm{t}}L_{\mathrm{r}}}}v_{\mathrm{t}} \tag{5.63}$$

The output current reference I_{bref} can still be converted to the transmitter-side current reference $|i_{\mathrm{pkref}}|$ using Equation (5.58), while the output voltage reference V_{bref} needs to be modified by substituting Equations (5.62) and (5.63) into Equation (5.59) and eliminating M as

$$|v_{\mathrm{tref}}| = \frac{4}{\pi}\sqrt{\frac{\omega_{\mathrm{CC\_LCLS}}L_{\mathrm{t}}L_{\mathrm{r}}|i_{\mathrm{r}}|}{|v_{\mathrm{t}}|}}\left(V_{\mathrm{bref}} + 2V_{\mathrm{D}}\right) \tag{5.64}$$

The simulated I_{b} and V_{b} for an LCL-S-compensated WPT system with the communication-free self-oscillating control are exhibited in Figure 5.28. The key system parameters are $L_{\mathrm{t}}=L_{\mathrm{r}}=L_{\mathrm{f}}=48.2$ μH, $C_{\mathrm{t}}=C_{\mathrm{r}}=52.5$ nF, $M=7.1$ μH, and $V_{\mathrm{dc}}=10$ V. Here, L_{f} is the compensated inductance at the transmitter side. The load resistance R_{L} toggles between 7.5 and 10 Ω at 0.04, 0.08, and 0.12 s. Operating with reference setpoints of 1.2 A (CC mode) and 1 V (CV mode), the system executes a controlled mode transition at $t=0.08$ s. The PI controller gains are tuned to $K_{\mathrm{p}}=30$ and $K_{\mathrm{i}}=40{,}000$ for CC regulation, and $K_{\mathrm{p}}=10$ and $K_{\mathrm{i}}=10{,}000$ for CV operation. Simulation results confirm precise tracking of both output

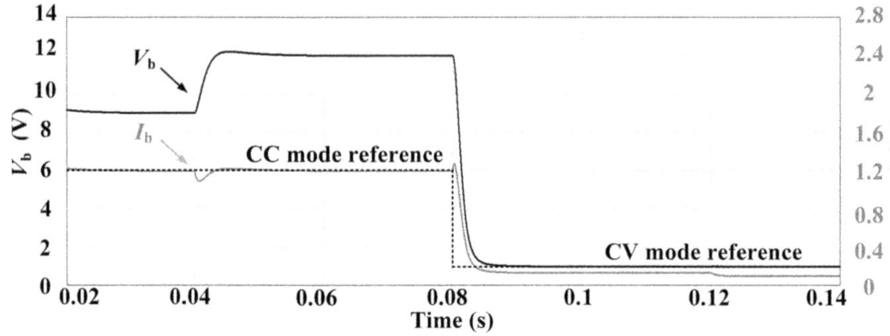

FIGURE 5.28
Simulated waveforms of V_b and I_b for the LCL-S-compensated WPT system with the communication-free self-oscillating control.

current and voltage to their reference values, even under dynamic load variations. These findings conclusively validate the effectiveness of the communication-free self-oscillating control strategy in LCL-S-compensated WPT systems.

The LCL-S compensation requires the compensated inductance to match the transmitter coil inductance, a condition often impractical in real-world applications. To address this limitation, LCC-S compensation was introduced, adding a series capacitance (i.e., C_f) alongside the transmitter coil to reduce the required inductance. As a result, LCC-S compensation is more widely adopted in practice than LCL-S. The ideal operating frequencies for achieving CC and CV outputs in LCC-S-compensated networks can be derived as

$$\omega_{CC\_LCCS} = \omega_o \sqrt{1 - \sqrt{\frac{L_f}{L_t}}} \tag{5.65}$$

$$\omega_{CV\_LCCS} = \omega_o \tag{5.66}$$

The ideal output current in CC mode and the ideal output voltage in CV mode can be further determined as

$$i_r = j\frac{\sqrt{L_t}\,v_t}{\omega_{CC\_LCCS} M \sqrt{L_f}} \tag{5.67}$$

$$v_r = \frac{M}{L_f} v_t \tag{5.68}$$

Figure 5.29 presents the simulated waveforms of the I_b and V_b for an LCC-S-compensated WPT system controlled by the proposed communication-free self-oscillating strategy. The key system parameters are as follows: $L_t = L_r = 48.2$ μH, $L_f = 24.2$ μH, $C_t = C_f = 104.7$ nF, $C_r = 52.5$ nF, $M = 7.2$ μH, and

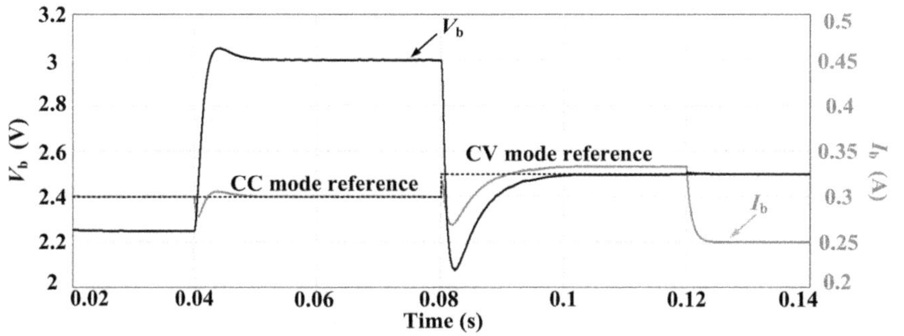

FIGURE 5.29
Simulated waveforms of V_b and I_b for the LCC-S-compensated WPT system with the communication-free self-oscillating control.

$V_{dc} = 10$ V. At $t = 0.04$ s, the load resistance R_L steps down from 10 to 7.5 Ω, then returns to 10 Ω at $t = 0.08$ s before repeating the 10 to 7.5 Ω transition at $t = 0.12$ s. The controller maintains 0.3 A in CC mode until $t = 0.08$ s, when it automatically switches to CV operation at 2.5 V. Controller parameters are set to $K_p = 50$ and $K_i = 40{,}000$ for CC mode, and $K_p = 100$ and $K_i = 40{,}000$ for CV mode operation. The simulation results demonstrate that the communication-free self-oscillating control scheme effectively regulates LCC-S-compensated WPT systems in both CC and CV operational modes.

5.2.2.4 Evaluation of the Communication-Free Self-Oscillating Control

The communication-free self-oscillating control offers significant advantages, mainly due to its simplicity and cost-effectiveness. These benefits stem not only from its intuitive algorithm, which can be easily implemented using low-cost digital controllers or simple analog circuits, but also from its straightforward pick-up coil design methodology. This approach is particularly advantageous for next-generation WPT systems operating at high or ultrahigh frequencies, where the performance requirements for sensing and signal-processing chips are substantially more demanding than those for low-frequency systems. Many of the transmitter-side control schemes discussed earlier face implementation challenges due to hardware limitations. In contrast, the communication-free self-oscillating control circumvents these issues entirely.

Nevertheless, the communication-free self-oscillating control faces several critical challenges. (i) Output offsets due to ESRs: The presence of ESRs can induce output offsets. While this issue has been addressed in two-stage WPT systems with different compensation networks [16] and domino WPT systems [17], the proposed solutions heavily rely on communication feedback of receiver-side parameters. One potential remedy is to replace out-of-band communication with in-band communication techniques for parameter feedback, though this would increase the complexity of the self-oscillating control itself. Alternatively, AI-based offset compensation could be explored, but no such studies have been conducted to date, leaving the validity of this approach unverified. Given these constraints, I recommend employing self-oscillating control primarily in WPT systems with high Q factors. For next-generation high-frequency WPT systems, however, this control method may exhibit greater adaptability. (ii) Limited application scope: The discussed communication-free self-oscillating control, which relies on a pick-up coil, necessitates series compensation on the receiver side. In practice, LCC compensation networks are more prevalent for high-power bidirectional applications. Additionally, the method requires prior knowledge of the mutual inductance between the transmitter and receiver coils to generate transmitter-side references, implying its suitability is largely restricted to static WPT scenarios. Consequently, the proposed control scheme may only be viable for static, low- to medium-power WPT systems, significantly narrowing its applicability. (iii) Dependence on coil coupling conditions: The efficacy of this control method hinges on two key coupling conditions: (1) decoupling between the transmitter coil and the pick-up coil, and (2) strong coupling between the receiver coil and the pick-up coil. Thus, meticulous design of the pick-up coil is paramount to ensure successful implementation.

5.2.3 Data-Driven Control Schemes

I anticipate that data-driven control schemes will dominate next-generation WPT systems once sufficient supporting hardware and training data become available. AI-based models could directly generate optimal control variables (such as operating frequency and phase shift angle) for transmitter-side driving circuits, eliminating the need for additional compensation mechanisms. My conceptual framework of data-driven control schemes for next-generation WPT systems is depicted in Figure 5.30. All critical receiver-side information, including circuit topology, load characteristics, and electrical, electromagnetic, thermal, and mechanical parameters, will be uploaded to the cloud. These data will be temporarily stored for training and validating physics-informed neural networks (PINNs). The cloud-based PINNs will then generate optimized control variables (e.g., operating frequency and phase shift angle) for local transmitter-side inverter drivers.

I think the proposed framework could offer three notable advantages:

i. PINNs versus neural networks:

PINNs are uniquely suited for creating digital twins of WPT systems, as they integrate physical laws into the learning process. This approach significantly reduces data storage and computational demands compared to conventional neural networks.

ii. Scalability via distributed data:

The cloud-based PINNs can aggregate real-time data from millions of WPT systems, enabling the digital twins to accurately model virtually all possible physical layer scenarios.

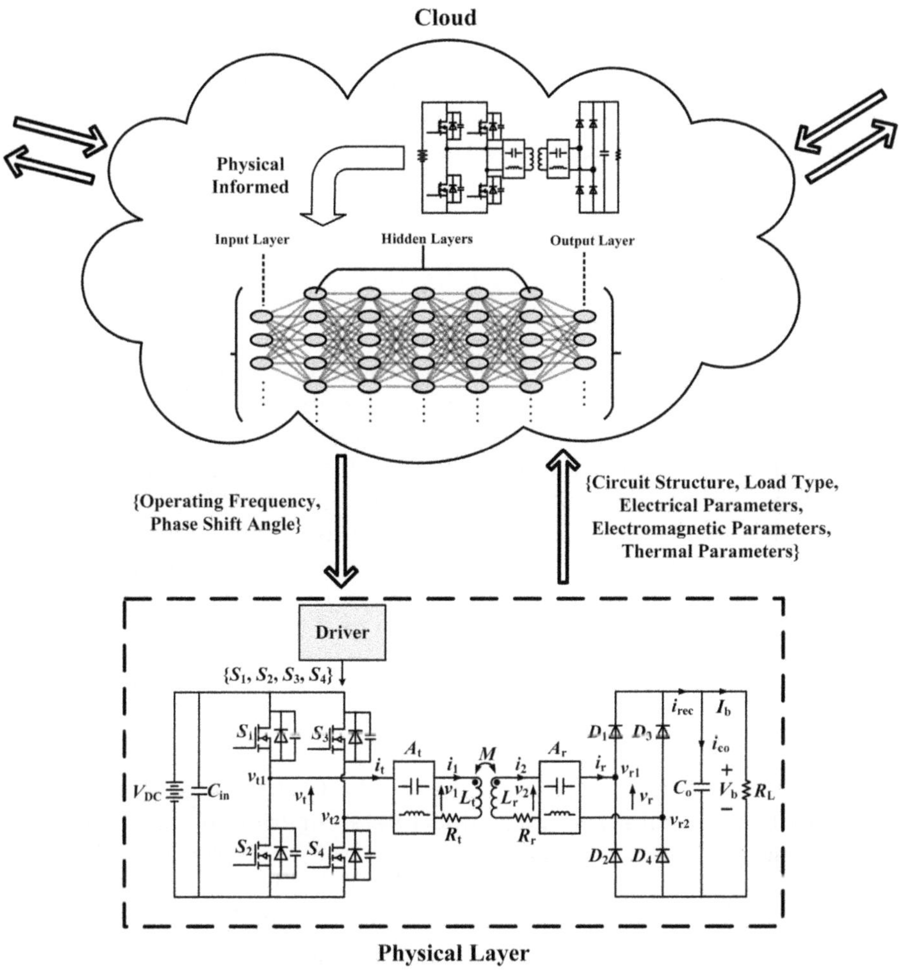

FIGURE 5.30
My conceptual framework of data-driven control schemes for next-generation WPT systems.

iii. Asynchronous and flexible communication:

1. Unlike conventional control schemes requiring real-time bidirectional communication, this method decouples data upload (receiver-to-cloud) and download (cloud-to-transmitter).
2. Edge computing can preprocess receiver data locally when cloud uploads are impractical (e.g., in low-connectivity environments).
3. Transmitter control commands can be delivered via wireless or wired communication (including in-band power-line communication).

Moreover, the control variables are applied to the transmitter-side driver circuitry with near-instantaneous response, eliminating the need for feedback delays or computational latency. As a result, WPT systems employing these data-driven control schemes exhibit dynamic performance characteristics comparable to open-loop controlled systems.

5.3 Control Algorithms

Next-generation WPT systems employing this transmitter-side control paradigm can utilize a wide variety of control algorithms, similar to the diversity seen in resonant converter controls. Potential classification schemes include: (i) linear versus nonlinear control, (ii) frequency-domain versus time-domain approaches, (iii) model-based versus model-free methods, etc. The model-based/model-free categorization proves most relevant for our discussion, as it directly addresses the fundamental challenge of parameter variations affecting model accuracy, a key issue in transmitter-side monitoring/control that has been a recurring theme in this book. This chapter will therefore focus on this particular classification.

5.3.1 Model-Based Control

Finite-control-set model predictive control (FCS-MPC) is a well-established sequential control strategy that offers superior dynamic response and enhanced disturbance robustness compared to conventional PI-based control in power electronics systems [18–21]. Due to its computational efficiency and effectiveness, FCS-MPC with one-step or two-step prediction horizons is widely employed for output voltage regulation in dynamic WPT systems [22–24]. In this chapter, two FCS-MPC control strategies are introduced: model predictive-based phase shift control (MPPC) and model predictive-based frequency control (MPFC) [11]. The presented MPPC and MPFC are representative FCS-MPC strategies employing one-step prediction. Both controllers are developed based on canonical system models and practical considerations, such as dead time, turn-on and turn-off delays, parasitic capacitances of the transmitter-side inverters, the bifurcation phenomenon inherent to IPT systems, and the determination of initial switching frequencies. Compared to their PI-based counterparts, the MPPC and MPFC offer superior dynamic performance, including reduced overshoot and faster settling times. However, the MPPC is susceptible to notching effects and current distortion caused by dead time. It also struggles to achieve soft switching under light load conditions. In contrast, the MPFC is inherently immune to dead time effects and reliably ensures soft switching of the transmitter-side inverters across a broad load range. As a result, the MPFC-controlled inverters exhibit lower EMI compared to those controlled by MPPC. Nevertheless, the accuracy of the MPFC may be influenced by the intrinsic frequency bifurcation characteristics of WPT systems. To address this, a frequency selection algorithm is proposed to determine the optimal initial switching frequency of the transmitter-side inverter. This optimal frequency effectively ensures ZVS, enhancing efficiency and reliability. Therefore, the proposed MPFC is more suitable than MPPC for the next-generation WPT systems, particularly when fixed-frequency operation is not a strict requirement.

5.3.1.1 Fundamental Analysis of the MPC for Transmitter-Side Regulation

For analytical clarity without loss of generality, an SS-compensated WPT system is selected for study. The circuit diagram of the investigated system is presented in the physical layer of Figure 5.30. Based on Fourier analysis, the fundamental components of v_t and i_r can be derived as

$$v_{t1}(t) = \frac{4V_{DC}\cos\left(\frac{\alpha}{2}\right)}{\pi}\sin(\omega t) \tag{5.69}$$

$$i_{r1}(t) = \frac{4V_{DC}\omega M\cos\left(\frac{\alpha}{2}\right)}{\pi\left|Z_t Z_r + \omega^2 M^2\right|}\sin\left(\omega t + \frac{\pi}{2} + \theta\right) \tag{5.70}$$

where $Z_t = j\omega L_t + \frac{1}{j\omega C_t} + R_t$, $Z_r = j\omega L_r + \frac{1}{j\omega C_r} + R_r + R_{eq}$, and θ is the phase angle of the term $Z_tZ_r+\omega^2M^2$, while the magnitude of $Z_tZ_r+\omega^2M^2$ (i.e., $|Z_tZ_r+\omega^2M^2|$) is constant in one cycle. When the MPPC is applied to the WPT system, the switching frequency is controlled at the resonant frequency, the impedances Z_t and Z_r are simplified to R_t and R_r+R_{eq}, respectively, and θ becomes zero. Based on Equation (5.70), the average output current of the diode-bridge rectifier can be calculated as

$$i_{rec}(t)_{T_s} \cong \frac{1}{T_s}\int_0^{T_s}\left|i_{r1}(t)\right|dt = \frac{16V_{DC}Mf\cos\left(\frac{\alpha}{2}\right)}{\pi\left|Z_t Z_r + \omega^2 M^2\right|} \tag{5.71}$$

where f is the switching frequency and T_s is the sampling period. The filter capacitor current i_{co} can be expressed as

$$i_{co}(t) = i_{rec}(t)_{T_s} - I_b(t)_{T_s} = C_o\frac{d\left(I_b(t)_{T_s}R_L\right)}{dt} \tag{5.72}$$

By applying Euler's forward discretization to Equation (5.71), the output current at the next sampling instant (i.e., $I_b[k+1]$) can be estimated from the current output $I_b[k]$ as

$$I_b[k+1] = \frac{16V_{DC}M\cos\left(\frac{\alpha}{2}\right)}{C_oR_L\pi\left|Z_tZ_r+\omega^2M^2\right|} + \left(1 - \frac{1}{C_oR_Lf}\right)I_b[k] \tag{5.73}$$

Clearly, the output current can be regulated through f and α. The proposed MPFC and MPPC are specifically designed to control f and α, respectively. In both MPC strategies, $I_b[k+1]$ is regulated to track the reference I_{bref} using a canonical cost function.

$$\min J = \left|I_{bref} - I_b[k+1]\right| \tag{5.74}$$

$$\text{s.t.} \quad 0 \le I_{bref} \le I_{brefmax} \tag{5.75}$$

$$f_{min} \le f \le f_{max} \tag{5.76}$$

The output current reference and switching frequency of the system are strictly constrained by the boundaries defined in Equations (5.75) and (5.76). The current constraints are imposed to ensure system stability, while the frequency constraints are determined by hardware limitations. For the MPFC, α is maintained at zero. At sampling time k, the output currents $I_b[k+1]$ are predicted for three frequency points: f, $f+\Delta f$, and $f-\Delta f$, where Δf represents the step change in switching frequency. The frequency corresponding to the minimum cost function value is then selected as the optimal switching frequency for the transmitter-side inverter at sampling time $k+1$. In the MPPC, following conventional phase shift control methods, the switching frequency f is fixed at the resonant frequency. The output current $I_b[k+1]$ varies monotonically with the phase shift angle α. By adjusting α with a step change $\Delta\alpha$, the predicted output current increases or decreases correspondingly. Similar to the MPFC, either a fixed or variable $\Delta\alpha$ can be used. The optimal phase shift angle for the transmitter-side inverter at sampling time $k+1$ is determined by comparing the cost functions of α, $\alpha+\Delta\alpha$, and $\alpha-\Delta\alpha$, selecting the value that minimizes the cost function.

5.3.1.2 Impact of Transmitter-Side Inverter Parasitic Parameters on MPPC Performance

To investigate the impact of key parasitic parameters in the transmitter-side inverter on MPPC performance, the inverter's operational states over one switching cycle are analyzed using the characteristic waveforms shown in

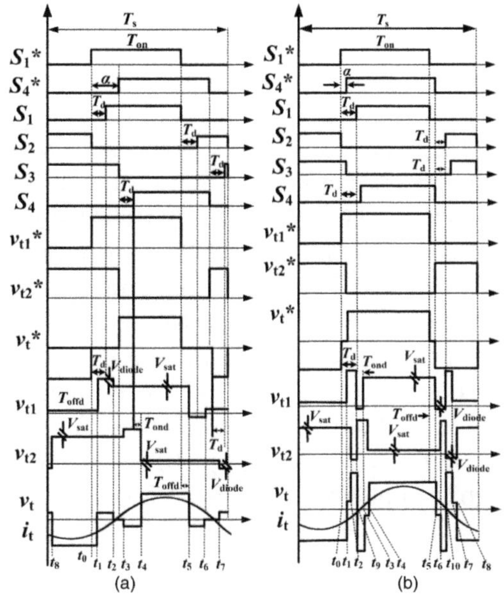

FIGURE 5.31
Main characteristic waveforms of the transmitter-side inverter controlled by MPPC with (a) $\alpha > \varphi_{Td+Tond-Toffd}$ and (b) $\alpha < \varphi_{Td+Tond-Toffd}$.

Figure 5.31. S_1^* and S_4^* represent ideal gate pulses without dead time, while S_1, S_2, S_3, and S_4 correspond to non-ideal gate pulses that include dead time. Similarly, v_{t1}^* and v_{t2}^* denote the ideal phase-leg voltages, and v_t is the ideal square-wave voltage between the two legs. v_{t1} and v_{t2} are the non-ideal phase-leg voltages, and v_t is the non-ideal square-wave voltage between the two legs. The parameters T_d, T_{ond}, and T_{offd} refer to the dead time, turn-on delay, and turn-off delay, respectively. The dead time angle is defined as φ_{Td}. The presence of dead time introduces notching effects [11]. When turn-on and turn-off delays are considered, the notching cut-off angle extends to $\varphi_{Td+Tond-Toffd}$, which corresponds to $\omega_s(T_d + T_{ond} - T_{offd})$. It is important to note that these cut-off angles may vary depending on the phase angle between v_{t1} and i_t (i.e., φ) and the phase angle between v_{t1} and v_t (i.e., φ_y). Additionally, V_{sat} and V_{diode} denote the forward voltage drops of the switching devices and body diodes, respectively.

The operating states of the inverter for $\alpha > \varphi_{Td+Tond-Toffd}$ (as illustrated in Figure 5.31a) are discussed below:

- Stage 1 (from t_0 to t_1): S_1^* is turned on, but the actual gate signal S_1 remains off due to dead time. Q_2 is not turned off immediately, and Q_3 remains on. As a result, i_t flows through Q_2 and Q_3. The phase voltage v_{t1} increases by V_{sat}, while v_{t2} decreases by V_{sat}. Consequently, the phase voltage difference v_t is reduced by $2V_{sat}$.

- Stage 2 (from t_1 to t_2): S_2 turns off, while S_3 remains on. i_t now flows through the bode diodes of S_1 and S_3. In this stage, v_{t1} increases by V_{diode}, while v_{t2} remains decreased by V_{sat}, resulting in an increase in v_t by $V_{sat}+V_{diode}$.
- Stage 3 (from t_2 to t_3): As i_t transitions from negative to positive, S_1 turns on while S_3 remains on. The current flows through S_1 and S_3. Both v_{t1} and v_{t2} decrease by V_{sat}, leading to a phase voltage difference $v_t=0$.
- Stage 4 (from t_3 to t_4): S_1 remains on while S_3 turns off. i_t flows through S_1 and the body diode of S_3. Here, v_{t1} remains decreased by V_{sat} and v_{t2} increases by V_{diode}. Thus, v_t increases by $V_{sat}+V_{diode}$.
- Stage 5 (from t_4 to t_5): S_4 is turned on while S_1 remains on. The current i_t flows through S_1 and S_4. In this case, v_{t1} decreases by V_{sat}, and v_{t2} increases by V_{sat}, resulting in a reduction of v_t by $2V_{sat}$.
- Stage 6 (from t_5 to t_6): S_1 is turned off, while S_4 remains turn-on state. i_t flows through the bode diodes of S_2 and S_4. The phase voltage $v_{t1}$is increased by V_{diode}, while the phase voltage v_{t2} is increased by V_{sat}. Therefore, v_t is increased by $V_{diode}+V_{sat}$.
- Stage 7 (from t_6 to t_7): As i_t changes from positive to negative, S_2 turns on while S_4 remains on. The current now flows through S_2 and S_4. Both v_{t1} and v_{t2} increase by V_{sat}, resulting in $v_t=0$.
- Stage 8 (from t_7 to t_8): S_4 turns off while S_2 remains on, and i_t flows through S_2 and the body diode of S_4. In this stage, v_{t1} increases by V_{sat} and v_{t2} increases by V_{diode}, giving a total increase in v_t of $V_{sat}+V_{diode}$.
- Stage 9 (from t_8 to t_0): S_3 turns on while S_2 remains on. With i_t being negative, the current flows through S_2 and S_3. The changes in v_{t1}, v_{t2}, and v_t are identical to those observed in Stage 1.

  For $\alpha<\varphi_{Td+Tond-Toffd}$, Stage 3 is replaced by Stages 10 and 11, and Stage 7 is modified to Stages 12 and 13. The remaining operation stages remain unchanged, as exhibited in Figure 5.31b.

- Stage 10 (from t_2 to t_9): S_3 is turned off, and the current i_t remains negative, flowing through the body diodes of S_1 and S_4. During this period, both v_{t1} and v_{t2} increase by V_{diode}, resulting in a phase voltage difference v_t increased by $2V_{diode}$.
- Stage 11 (from t_9 to t_3): As i_t transitions from negative to positive, it flows through the body diodes of S_2 and S_3. Again, both v_{t1} and v_{t2} increase by V_{diode}, causing v_t to increase by $2V_{diode}$.
- Stage 12 (from t_6 to t_{10}): S_4 is turned off, and the positive current i_t flows through the body diodes of S_2 and S_3. Both phase voltages v_{t1} and v_{t2} increase by V_{diode}, leading to a phase voltage difference v_t of $2V_{diode}$.
- Stage 13 (from t_{10} to t_7): As i_t changes from positive to negative, it flows through the body diodes of S_1 and S_4. Again, v_{t1} and v_{t2} increase by V_{diode}, resulting in a phase voltage difference v_t of $2V_{diode}$.

By comparing the waveforms of v_t^* and v_t in Figure 5.31, it is evident that the duration of the positive v_t is reduced when T_d, T_{ond}, and T_{offd} are taken into account. For an ideal system employing the proposed MPPC, the phase shift angle α and the sensitivity of I_b with respect to α can be derived from Equation (5.73) as follows:

$$\alpha = 2\arccos\left(\frac{I_b[k]\pi\left|Z_t Z_r + \omega^2 M^2\right|}{16 V_{DC} M f}\right) \tag{5.77}$$

and

$$\left|\frac{\partial I_b[k]}{\partial \alpha}\right| = \frac{8V_{DC} f}{\pi\left|\frac{R_t\left(R_r + R_{eq}\right)}{M} + \omega^2 M\right|}\sin\left(\frac{\alpha}{2}\right) \tag{5.78}$$

It is evident that the phase shift angle α tends to approach zero under heavy load conditions, particularly when V_{DC} is reduced and R_L is increased. Conversely, under light load conditions, α remains non-zero. Moreover, deviations in α can lead to output current offsets. Due to the presence of T_d, T_{ond}, and T_{offd}, the value of α calculated from Equation (5.77) deviates from the actual α, resulting in a reduced output range. Even when $\alpha < \varphi_{Td+Tond-Toffd}$, notching effects in v_t remain unavoidable. To explicitly highlight the impact of parasitic parameters on the MPC performance, the absolute voltage difference between v_t and v_t^* during the positive half-cycle (referred to as the voltage loss Δv_t) is evaluated. A higher Δv_t indicates a greater discrepancy between the ideal and actual voltages, signifying more severe degradation. For the MPPC, the average value of v_t^* over the positive half-cycle is

$$v_t^* = \frac{\pi - \alpha}{\pi} V_{DC} \tag{5.79}$$

The average value of v_t during the positive half-cycle is given by

$$v_t = \left[\frac{\pi - \alpha}{\pi} - \frac{2(T_d + T_{ond} - T_{offd})}{T_s}\right](V_{DC} - 2V_{sat}) \tag{5.80}$$

Then, the voltage loss Δv_t is defined as

$$\Delta v_t = \left[\frac{4(T_d + T_{ond} - T_{offd})}{T_s} - \frac{2\pi - 2\alpha}{\pi}\right]V_{sat} - \frac{2(T_d + T_{ond} - T_{offd})}{T_s} V_{DC} \tag{5.81}$$

Clearly, Δv_t is influenced by V_{DC}, T_d, T_{ond}, T_{offd}, V_{sat}, and α. Among these, T_{ond}, T_{offd}, and V_{sat} are primarily determined by the characteristics of the gate driver circuitry and the switching devices, while T_d and α can be regulated through software. For active switches with negligible V_{sat}, the voltage loss expression in Equation (5.81) can be simplified as

$$\Delta v_t = \frac{2(T_d + T_{ond} - T_{offd})}{T_s} V_{DC} \tag{5.82}$$

It is evident that dead time, turn-on delay time, and turn-off delay time significantly impact the regulation performance of the MPPC. Furthermore, as indicated by Equation (5.78), the sensitivity of I_b with respect to α decreases monotonically across all operating conditions. This implies that the regulation of I_b becomes less susceptible to disturbances in α as α approaches zero. For a fixed α, the sensitivity further decreases when the load resistance R_L increases, the input voltage V_{DC} decreases, or the mutual inductance M increases, particularly when $M > \sqrt{\frac{R_t(R_r + R_{eq})}{\omega^2}}$. Consequently, the MPPC offers more robust output current regulation in WPT systems operating under lower input voltage, stronger coupling, and heavier load conditions. Moreover, it is well known that the parasitic capacitances of active switches can lead to hard switching in WPT systems under light load conditions [25]. Therefore, it is essential to either compensate for the dead time within the MPPC or incorporate auxiliary circuits to ensure soft switching of the transmitter-side inverter.

5.3.1.3 Impact of Transmitter-Side Inverter Parasitic Parameters on MPFC Performance

For MPFC, the transmitter-side inverter can operate in nine states. The characteristic waveforms are shown in Figure 5.32.

For $\varphi > \frac{\varphi_{T_d + T_{ond} - T_{offd}}}{2}$, as shown in Figure 5.32a,

- Stage 1 (from t_0 to t_1): S_1^* is turned on, but S_1 remains in the off state. S_2 and S_3 are not turned off immediately, allowing i_t to flow through them. As a result, v_{t1} increases by V_{sat}, while v_{t2} decreases by V_{sat}, leading to a net reduction in the phase voltage difference v_t by $2V_{sat}$.
- Stage 2 (from t_1 to t_2): S_2 and S_3 are turned off, i_t now flows through the body diodes of S_1 and S_4. Both v_{t1} and v_{t2} increase by V_{diode}, resulting in a total increase of v_t by $2V_{diode}$.
- Stage 3 (from t_2 to t_3): S_1 and S_4 are turned on, allowing i_t to flow through them. v_{t1} decreases by V_{sat}, while v_{t2} increases by the same amount, reducing v_t by $2V_{sat}$.
- Stage 4 (from t_3 to t_4): S_1 and S_4 are turned off. i_t flows through the body diodes of S_2 and S_3, causing both v_{t1} and v_{t2} to increase by V_{diode}. As a result, v_t increases by $2V_{diode}$.

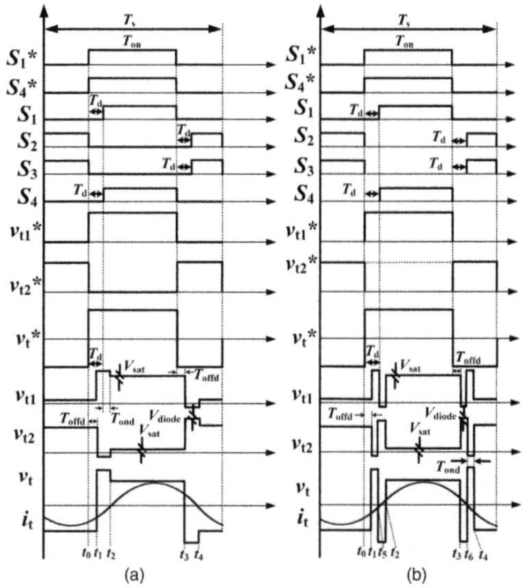

FIGURE 5.32
Main characteristic waveforms of the transmitter-side inverter controlled by MPFC with (a) $\varphi > \frac{\varphi_{T_d+T_{ond}-T_{offd}}}{2}$ and (b) $\varphi < \frac{\varphi_{T_d+T_{ond}-T_{offd}}}{2}$.

- Stage 5 (from t_4 to t_0): S_2 and S_3 are turned on again, and i_t resumes flowing through them. The variations in v_{t1}, v_{t2}, and v_t mirror those observed in Stage 1.

  For $\varphi < \frac{\varphi_{T_d+T_{ond}-T_{offd}}}{2}$, as shown in Figure 5.32b, Stage 2 is replaced by Stage 6 and Stage 7, while Stage 4 is substituted by Stage 8 and Stage 9.

- Stage 6 (from t_1 to t_2): The changes in v_{t1}, v_{t2}, and v_t are identical to those in Stage 2.
- Stage 7 (from t_5 to t_2): As i_t transitions from negative to positive, it flows through the body diodes of S_2 and S_3. Both v_{t1} and v_{t2} increase by V_{diode}, resulting in an increase of v_t by $2V_{diode}$.
- Stage 8 (from t_5 to t_2): The changes in v_{t1}, v_{t2}, and v_t are identical to those in Stage 4.
- Stage 9 (from t_6 to t_4): As i_t transitions from positive to negative, it flows through the body diodes of S_1 and S_4. Both v_{t1} and v_{t2} increase by V_{diode}, leading to an increase in v_t by $2V_{diode}$.

By comparing the schematic waveforms of v_t^* and v_t in Figure 5.32a, it is evident that the duration of the positive v_t remains unchanged relative to v_t^*, even when T_d, T_{ond}, and T_{offd} are taken into account. Notably, the period of positive v_t and the voltage loss Δv_t observed in Figure 5.32b are comparable to those in Figure 5.31b, as the operating frequency exceeds the resonant frequency. Therefore, only the voltage loss depicted in Figure 5.32a is analyzed. In this case, the average v_t^* during the positive half-cycle for MPFC is equal to V_{DC}, while the average v_t in the same interval is

$$v_t = V_{DC} - 2V_{sat} + \frac{4(T_d + T_{ond} - T_{offd})}{T_s}(V_{sat} + V_{diode}) \tag{5.83}$$

The voltage loss Δv_t can be determined as

$$\Delta v_t = -\left[2 - \frac{4(T_d + T_{ond} - T_{offd})}{T_s}\right]V_{sat} + \frac{4(T_d + T_{ond} - T_{offd})}{T_s}V_{diode} \tag{5.84}$$

It can be seen from Equation (5.84) that the voltage loss is influenced by T_d, T_{ond}, T_{offd}, V_{sat}, and V_{diode}. For a given set of gate driving circuits and switching devices, T_{ond}, T_{offd}, V_{sat}, and V_{diode} are fixed, leaving T_d as the only adjustable parameter. If both V_{sat} and V_{diode} are negligible, the voltage loss Δv_t approaches zero, indicating that the impact of dead time and switching delays on MPFC regulation is minimal. Furthermore, soft switching in the transmitter-side inverter under MPFC can be more readily achieved by setting the switching frequency higher than the resonant frequency [26]. A comparison of the effects of parasitic parameters between MPPC and MPFC is presented in Table 5.5. Clearly, MPFC is less susceptible to the adverse effects of transmitter-side inverter parasitic components than MPPC.

5.3.1.4 Implementation Details of MPC

The control block diagram of the MPPC is shown in Figure 5.33a. At the next sampling time $k+1$, the output current is predicted using Equation (5.73), based on the calculated phase shift angle $\alpha[k+1]$ and the measured output current $I_b[k]$. This predicted output current is then controlled to track the reference current I_{bref}, using the objective function and constraints outlined in Equations (5.74), (5.75), and (5.76). The control output is the optimal phase shift angle (i.e., α_{opt}) which is applied to the transmitter inverter of the WPT system and is adjusted by a step change $\Delta\alpha$ for $\alpha[k+1]$.

TABLE 5.5

Main Differences of Parasitic Parameter Effects between MPPC and MPFC

Characteristics	MPPC	MPFC
Period of positive v_t as compared to that of v_t^*	Smaller	Same
Main effects on the voltage loss Δv_t	T_d, T_{ond}, T_{offd}, α, V_{DC}	T_d, T_{ond}, T_{offd}
Cut-off angle of notching effects	Random	Constant
Soft switching	Need compensation	No change

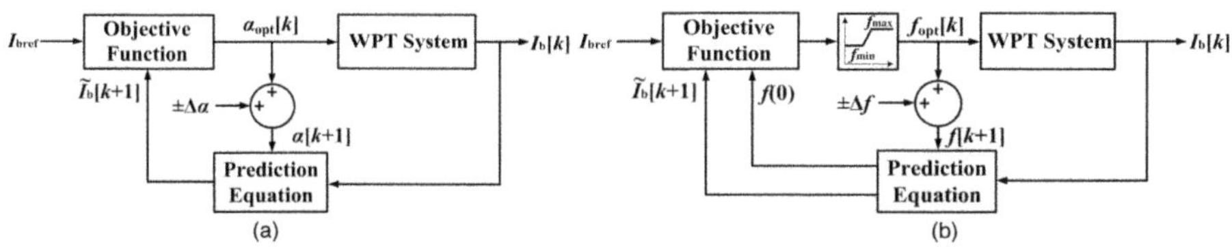

FIGURE 5.33
Control block diagrams of the proposed (a) MPPC and (b) MPFC for WPT systems.

For the MPFC, the control block diagram is shown in Figure 5.33b. The output current at the next sampling time $k+1$ is predicted using Equation (5.73), based on the calculated switching frequency $f[k+1]$ and the measured output current $I_b[k]$. Similar to the MPPC, the estimated output current is adjusted to track the reference using the objective function and constraints in Equations (5.74), (5.75), and (5.76). However, the control output in this case is the optimal switching frequency (i.e., f_{opt}) which is constrained within the lower bound f_{min} and upper bound f_{max}. The optimal frequency is applied to the WPT system and can be increased or decreased by a step change Δf for $f[k+1]$. The initial frequency of the MPFC (i.e., $f(0)$) is determined using Equation (5.73), with $I_b[k+1]$ set equal to $I_b[k]$ and both equal to the reference current I_{bref}. The flowchart illustrating the algorithm for determining $f(0)$ is shown in Figure 5.34.

Initially, the magnitude of the system transconductance g_m at the resonant frequency is computed using

$$g_m = \frac{\dot{I}_r}{\dot{V}_t} = \frac{j\omega M}{Z_t Z_r + \omega^2 M^2} \tag{5.85}$$

The operating frequency to achieve ZVS for the transmitter-side inverter (i.e., f_{zvs}) is determined accordingly as

$$f_{zvs} = \frac{\omega_o}{2\pi}\sqrt{\frac{\left(2 - \dfrac{1}{Q_r^2}\right) + \sqrt{\dfrac{4M^2}{L_t L_r} - \dfrac{4}{Q_r^2} + \dfrac{1}{Q_r^4}}}{2\left(1 - \dfrac{M^2}{L_t L_r}\right)}} \tag{5.86}$$

$$\text{s.t.} \quad \frac{4M^2}{L_t L_r} - \frac{4}{Q_r^2} + \frac{1}{Q_r^4} > 0 \tag{5.87}$$

where Q_r is the quality factor of the receiver resonator, determined by

$$Q_r = \frac{\omega_o L_r}{R_r + R_{eq}} \tag{5.88}$$

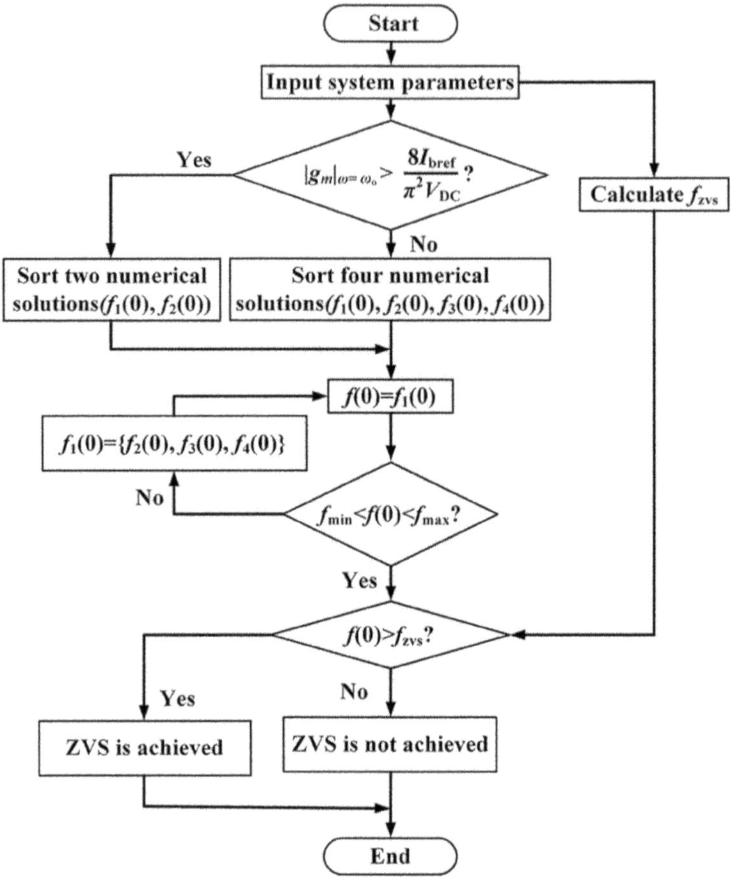

FIGURE 5.34
Flowchart of the algorithm to determine the initial frequency.

The transconductance g_m is compared to $\frac{8I_{bref}}{\pi^2 V_{DC}}$. If g_m exceeds this threshold, the WPT system exhibits two independent frequencies (i.e., $f_1(0) > f_2(0)$). Otherwise, it operates with four independent frequencies (i.e., $f_1(0) > f_2(0) > f_3(0) > f_4(0)$) [27]. To ensure ZVS, the initial frequency $f(0)$ is primarily assigned as $f_1(0)$. However, $f(0)$ is first validated against the frequency boundaries. If it lies outside these limits, $f(0)$ is reassigned to $f_2(0)$ (with analogous checks for $f_3(0)$ and $f_4(0)$). If within bounds, $f(0)$ is compared to f_{zvs} to determine whether ZVS operation is achievable.

Both MPPC and MPFC can be implemented cost-effectively using inexpensive digital controllers. However, MPFC requires additional computational effort to determine the initial frequency, whereas it achieves ZVS more easily through efficient software-based methods.

5.3.1.5 Experimental Verification

Experiments are performed on a prototype system with key parameters listed in Table 5.6. The nominal mutual inductance and load resistance are 6.243 μH and 1 Ω, respectively. The digital controller operates with a timer interrupt interval of 30 μs. Both the MPPC and MPFC algorithms are sampled at a frequency of 100 kHz. The inverter's turn-on and turn-off delays are 54 and 28 ns, respectively.

In the first case study, the load resistance R_L is varied from 1 to 7 Ω to evaluate the first dynamic response, while the output current reference I_{bref} is stepped from 0.84 to 1.68 A for the second. The dead time for MPPC is set to $T_d=0.1$ μs. The execution times for the proposed MPPC and MPFC algorithms are 21 and 28 μs, respectively. The initial switching frequency of MPFC is $f(0)=112.6$ kHz, calculated using Equation (5.77). The step sizes for MPPC and MPFC are $\Delta\alpha=0.018°$ and $\Delta f=0.25$ Hz, respectively. Figure 5.35 shows the waveforms of v_t, i_t, I_b, and V_b. For MPPC, the phase shift angle α stabilizes at 148°, 146°, and 110° across three steady states. For MPFC, the switching frequency f settles at 111.1, 108.7, and 102 kHz, respectively. In both methods, I_b closely tracks I_{bref} with minimal steady-state error. The average values $\overline{I_b}$ and $\overline{V_b}$ represent the mean of I_b and V_b, respectively. Both MPPC and MPFC demonstrate negligible overshoot during the first dynamic and maintain zero

TABLE 5.6

Key Parameters of the WPT System with MPPC and MPFC

Parameter	Values
Transmitter and receiver inductances L_t, L_r	60.9 μH
Transmitter and receiver capacitance C_t, C_r	41.59 nF
Transmitter and receiver resistance R_t, R_r	0.14 Ω
Input DC voltage V_{DC}	15 V
Output capacitance C_o	100 μF

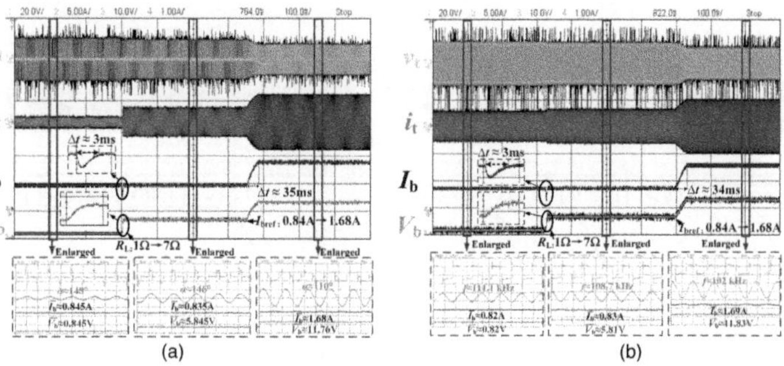

FIGURE 5.35
Waveforms of v_t, i_t, I_b, and V_b for the WPT system with (a) MPPC and (b) MPFC under R_L and I_{bref} variations. (Adapted from reference [11].)

overshoot during the second, even without retuning control parameters. The settling times for MPPC and MPFC are 35 and 34 ms, respectively, confirming that both strategies deliver fast and stable dynamic performance.

In the second case study, experiments are conducted on the WPT system with $V_{DC}=10$ V, $M=6.243$ μH, $R_L=1$ Ω, and $I_{bref}=1.72$ A to examine the impact of transmitter-side inverter dead time on MPPC performance. The dead time T_d is set to 0.5 μs. Figure 5.36 shows the waveforms of v_{t1}, v_{t2}, v_t, i_t, and I_b with and without dead time. The phase shift angle is fixed at 66° to ensure accurate tracking of I_{bref}. With $T_d=0.5$ μs, notching effects appear on v_t. As shown in the FFT spectra in Figure 5.37, the fundamental, third, and fifth harmonics (i.e., v_{tH1}, v_{tH3}, and v_{tH5}) remain nearly unchanged, while the seventh and ninth harmonics (i.e., v_{tH7} and v_{tH9}) increase notably. The amplitude of v_{tH7} increases from 0.14 to 0.47 V, while the amplitude of v_{tH9} rises from 0.51 to 0.73 V. The total harmonic distortion (THD) of v_t increases from 9.16% without dead time to 9.88% with dead time.

Additional experiments are conducted on the same WPT system using MPFC. The waveforms of v_{t1}, v_{t2}, v_t, i_t, and I_b are shown in Figure 5.38. By maintaining the switching frequency at 108.3 kHz, I_b is accurately regulated to track I_{bref} for both MPFC with and without dead time. Setting the transmitter-side inverter dead time to $T_d=0.5$ μs proves to be a reasonable choice,

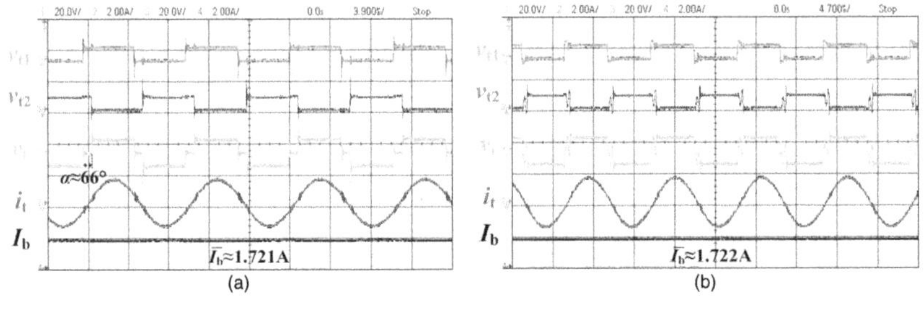

FIGURE 5.36
Waveforms of v_{t1}, v_{t2}, v_t, i_t, and I_b for the WPT system with MPPC: (a) without dead time and (b) with $T_d=0.5$ μs. (Adapted from reference [11]).

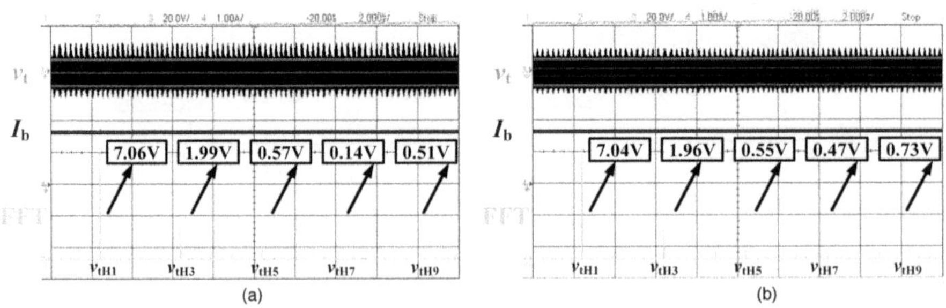

FIGURE 5.37
FFT spectra of v_t for the WPT system with MPPC: (a) without dead time and (b) with $T_d=0.5$ μs. (Adapted from reference [11].)

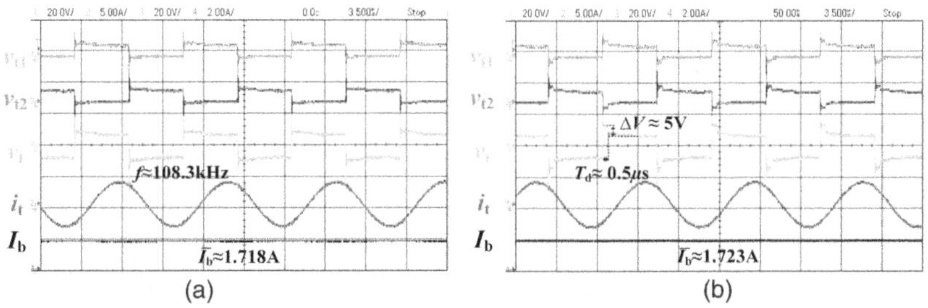

FIGURE 5.38
Waveforms of v_{t1}, v_{t2}, v_t, i_t, and I_b for the WPT system with MPFC: (a) without dead time and (b) with $T_d = 0.5$ μs. (Adapted from reference [11].)

as no notching effects (as seen in Figure 5.36b for MPPC) are observed. Furthermore, the phase difference between v_t and i_t remains consistent at 64° in both cases. These results confirm the analysis that dead time affects MPPC performance but has negligible impact on MPFC.

The EMI spectra of the transmitter-side inverter under MPPC and MPFC control are also measured. The comparative results are presented in Figure 5.39. While the peak radiation noise levels for both control methods remain below the EN55022B limit, MPFC exhibits noticeably lower emissions than MPPC in the 1–10 MHz range. Specifically, the peak noise levels of MPFC are 22 dBμV at 4 MHz and 13 dBμV at 5 MHz, compared to 35 and 27 dBμV at the same frequencies for MPPC.

5.3.1.6 Evaluation of the Model-Based Control

The MPPC and MPFC methods presented in [11] demonstrate superior dynamic performance. MPFC offers advantages in achieving soft-switching operation for the transmitter-side inverter, while MPPC may yield higher resonator efficiency compared to MPFC. Both MPC schemes require certain system parameters, such as mutual inductance, as well as output signals fed back from the receiver to the transmitter side. However, the transmitter-side monitoring schemes previously discussed can effectively mitigate these dependencies. The primary concern lies in potential parameter drifts over prolonged operation, which may lead to performance degradation or steady-state offsets. As such, periodic calibration is recommended to maintain optimal performance.

5.3.2 Model-Free Control

Some researchers suggest that model-free control represents a novel category distinct from model-based control. However, this distinction may be misleading. Many traditional control strategies, such as PI control and hysteresis control, are inherently model-free in the context of power electronics systems.

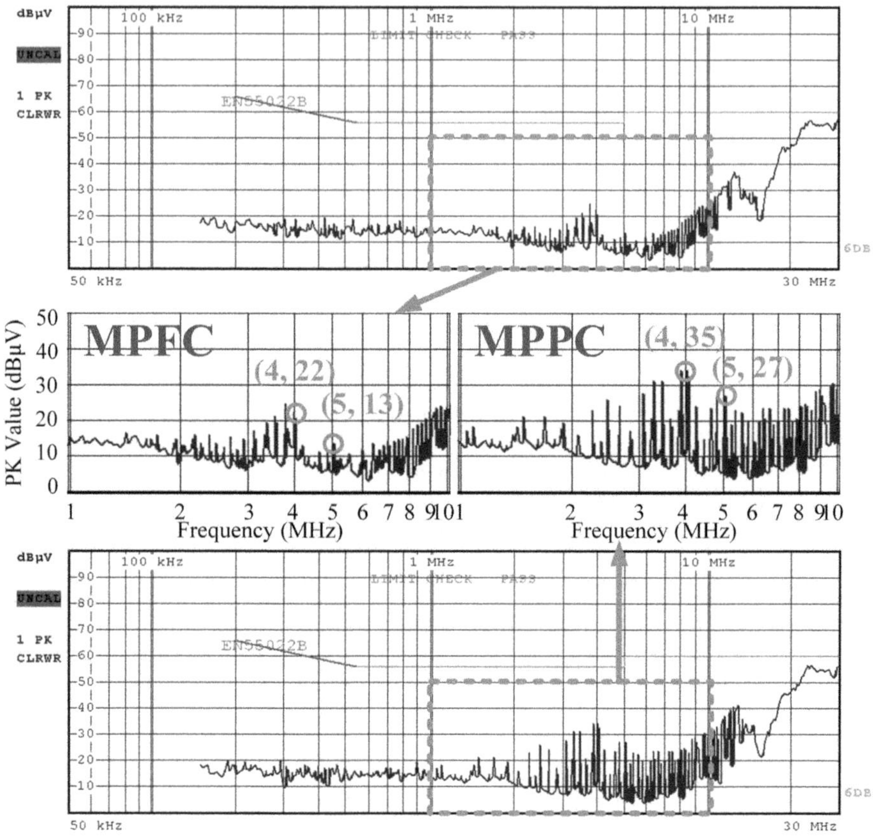

FIGURE 5.39
Comparative EMI spectra of the transmitter-side inverter with MPPC and MPFC. (Adapted from reference [11].)

Even predictive control does not necessarily imply the use of MPC, as various forms of prediction-based schemes exist without relying on full system models. However, it is important to clarify that the term "model" in this context refers specifically to a full system model derived from circuit topology and fundamental circuit laws. Models obtained through AI training or based on heuristic structures (such as the linear PI form of conventional PI) are not considered.

Conventional PI control is a typical model-free control, well-suited for linear systems or nonlinear systems operating within a limited range. Proportional and integral actions are often sufficient to eliminate steady-state errors and shape the dynamic response in power electronic systems. Derivative terms are generally avoided, as they can introduce chattering. However, the key limitation of conventional PI control lies in its poor robustness against disturbances and degraded performance when operating far from the nominal point, especially in power electronics, which involve switching behavior. This can be intuitively understood using a highly nonlinear curve with multiple local optima: a PI controller is typically tuned for one operating point and performs best along its local slope. When the system shifts to another region with a different slope, the controller becomes suboptimal unless its parameters are retuned. Adaptive PI controllers are built upon this principle, but they require additional feedback and lose the simplicity of linear control.

The next-generation WPT systems under investigation exhibit strong nonlinear characteristics, making conventional PI control less effective in terms of dynamic performance compared to the MPC discussed in Section 5.3.1. Figure 5.40 presents the experimental waveforms of v_t, i_t, I_b, and V_b for the WPT system controlled by conventional PI-based phase shift and frequency control. The first dynamic scenario involves a change in load resistance R_L from 1 to 7 Ω, while the second involves a step change in reference current I_{bref} from 0.84 to 1.68 A. The main specifications of the WPT system are given in Table 5.6. The occupation time of conventional PI-based phase shift control and frequency control are 3 and 28 μs, respectively. The proportional and integral gains of the PI-based phase shift and frequency control are optimally tuned to 3/A and 0.000025/A, and 1 Hz/A and 0.0004 Hz/A, respectively, to ensure that the settling time for the first dynamic change (i.e., 3 ms) matches that of the MPPC and MPFC. For the PI-based phase shift control, the phase shift angle α is regulated to 148°, 146°, and 110° at the three steady states, consistent with the MPPC. Similarly, for the PI-based frequency control, the operating frequency is maintained at 111.1, 108.7, and 102 kHz, matching the values used in the MPFC. Consequently, I_b is accurately regulated to track I_{bref} under both PI-based phase shift and frequency control. During the first dynamic response (R_L step change from 1 to 7 Ω), the observed overshoots are 2 V for the PI-based phase shift control and 5 V for the PI-based frequency control. During the second dynamic response (I_{ref} step increase from 0.84 to

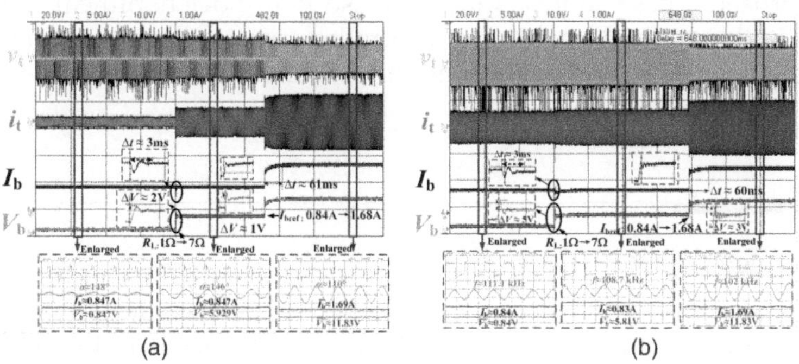

FIGURE 5.40
Waveforms of v_t, i_t, I_b, and V_b for the WPT system with (a) conventional PI-based phase shift control and (b) conventional PI-based frequency control under R_L and I_{bref} variations. (Adapted from reference [11]).

1.68 A), the PI-based phase shift control achieves 1 V overshoot and 61 ms settling time, while the PI-based frequency control achieves 3 V overshoot and 60 ms settling time. Obviously, conventional PI-based phase shift control and PI-based frequency control show markedly inferior dynamic performance compared to the MPC-based counterparts, particularly in overshoot suppression and response speed.

Therefore, it is essential to develop advanced model-free control strategies for the WPT system to achieve performance comparable to that of model-based approaches. The proportional resonant (PR) control strategy is an advanced model-free approach that offers infinite gain at a specific operating frequency, in contrast to the conventional PI control strategy, which provides infinite gain at DC. While practical implementations of PR control may not achieve true infinite gain, they can be designed as quasi-PR controllers to deliver very high gain at the target frequency. This capability helps significantly reduce steady-state error at that frequency and enhances dynamic performance. Another advanced model-free control technique is sliding mode control (SMC), which can be designed based on the large-signal model of the system. Certain forms of SMC combine the benefits of both hysteresis control and conventional PI control, offering robust performance under varying conditions. However, SMC often requires additional sensing and may introduce undesirable chattering effects. Some variants of SMC share a similar structure with adaptive PI control, where the proportional and integral gains are dynamically adjusted in real time based on feedback from certain measured parameters. If such feedback is independent of a predefined system model, these control strategies (whether SMC or adaptive PI) can be considered model-free. Therefore, leveraging AI to implement this type of adaptive feedback appears to be a highly promising model-free approach, especially given the significant advantages of AI highlighted in the previous chapter. Certainly, using AI to adjust other parameters (such as control references to form virtual references [28]) while keeping the PI controller's tuning parameters fixed is an alternative approach. These control strategies also fall under nonlinear model-free control and essentially adapt the controller structure to accommodate the changing operating conditions of nonlinear control systems.

References

1. Y. Jiang, J. Yang, J. Wu, K. Li, Y. Yang, and S. -C. Tan, "A secure fast charging control based on a machine learning-aided electrothermal model for Lithium-ion batteries," in *IEEE Energy Conversion Congress and Exposition (ECCE)*, Phoenix, AZ, USA, 2024, pp. 1873–1880.
2. S. Y. R. Hui and Y. Yang, "Battery charging system and method using dynamically adjusted battery voltage threshold for switching charging modes," U.S. Patent US12113390B2, Oct. 8, 2024.

3. H. -W. R. Liang, Y. Yang, and S. Y. R. Hui, "Improvement of lithium-ion battery charging from the state-of-the-art industrial JEITA guidelines to a hybrid temperature-regulated current control," *IEEE Trans. Power Electron.*, vol. 37, no. 6, pp. 6412–6423, Jun. 2022.
4. H. W. R. Liang, Y. Yang, L. He, J. Qu, C. K. Lee, and S. Y. R. Hui, "A multi-hysteresis control for minimizing battery charging time within industrial JEITA guidelines," *IEEE Trans. Ind. Electron.*, vol. 70, no. 8, pp. 8416–8425, Aug. 2023.
5. J. Yin, D. Lin, T. Parisini, and S. Y. Hui, "Front-end monitoring of the mutual inductance and load resistance in a series–series compensated wireless power transfer system," *IEEE Trans. Power Electron.*, vol. 31, no. 10, pp. 7339–7352, Oct. 2016.
6. K. Wang, J. Zeng, Y. Yang, and S. Y. R. Hui, "A general maximum energy efficiency tracking scheme for domino wireless power transfer systems with quasi-load-independent outputs," *IEEE Trans. Power Electron.*, vol. 39, no. 1, pp. 1840–1852, Jan. 2024.
7. Y. Yang, J. Zeng, and S. Y. R. Hui, "A fast primary-side current and voltage control for direct wireless battery chargers," *IEEE J. Emerg. Sel. Topics in Power Electron.*, vol. 12, no. 1, pp. 355–366, Feb. 2024.
8. Qi Specification [Online]: https://www.wirelesspowerconsortium.com/knowledge-base/specifications/download-the-qi-specifications/.
9. Y. Mao, Y. Yang, and K. Wang, "A primary-side fixed-frequency CC and CV output control for single-stage wireless battery chargers with series compensation receivers," *IEEE Trans. Transp. Electrif.*, vol. 10, no. 2, pp. 3407–3415, Jun. 2024.
10. L. Xu, Q. Chen, X. Ren, S. Wong, and C. K. Tse, "Self-oscillating resonant converter with contactless power transfer and integrated current sensing transformer," *IEEE Trans. Power Electron.*, vol. 32, no. 6, pp. 4839–4851, Jun. 2017.
11. H. Xiao, Y. Yang, K. Wang, and J. Wu, "Comparative studies of front-end model predictive control for direct inductive power transfer systems," *IEEE Trans. Power Electron.*, vol. 38, no. 10, pp. 11885–11897, Oct. 2023.
12. K. Yan, Q. Chen, J. Hou, X. Ren, and X. Ruan, "Self-oscillating contactless resonant converter with phase detection contactless current transformer," *IEEE Trans. Power Electron.*, vol. 29, no. 8, pp. 4438–4449, Aug. 2014.
13. L. Xu, G. Ke, Q. Chen, X. Ren, and Z. Zhang, "A self-oscillating resonant converter with precise output voltage control," in *IEEE International Power Electronics and Motion Control Conference (IPEMC2020-ECCE*, Nanjing, China, 2020, pp. 2946–2951.
14. X. Ren, Q. Chen, L. Cao, X. Ruan, S C. Wong, and C. K. Tse, "Characterization and control of self-oscillating contactless resonant converter with fixed voltage gain," in *International Conference on Power Electronics and Motion Control (IPEMC)*, Harbin, China, 2012, pp. 1822–1827.
15. K. Wang, Y. Yang, H. Wang, and E. K. -W. Cheng, "Frequency regulation scheme for double-sided LCC compensated inductive power transfer systems with quasi-load-independent outputs," in *IEEE Energy Conversion Congress and Exposition (ECCE)*, Nashville, TN, USA, 2023, pp. 6383–6388.
16. Z. Sun, K. Wang, Y. Wang, Y. Yang, S. -C. Tan, and S. Y. R. Hui, "Unveiling the essential output deviation factors of non-ideal inductive power transfer systems with self-oscillating control," in *2024 IEEE Energy Conversion Congress and Exposition (ECCE)*, Phoenix, AZ, USA, 2024, pp. 2169–2175.

17. K. Wang, J. Zeng, Y. Yang, and S. Y. R. Hui, "A general maximum energy efficiency tracking scheme for domino wireless power transfer systems with quasi-load-independent outputs," *IEEE Trans Power Electron.*, vol. 39, no. 1, pp. 1840–1852, Jan. 2024.
18. R. P. Aquilera, P. Lezana, and D. E. Quevedo, "Finite-control-set model predictive control with improved steady-state performance," *IEEE Tran. Ind. Informat.*, vol. 9, no. 2, pp. 658–667, May 2013.
19. J. Rodríguez and P. Cortés, *Predictive Control of Power Converters and Electrical Drives*. Hoboken, NJ: Wiley, 2012.
20. Y. Yang, S. -C. Tan, and S. Y. R. Hui, "Adaptive reference model predictive control with improved performance for voltage-source inverters," *IEEE Trans. Control Syst. Technol.*, vol. 26, no. 2, pp. 724–731, Mar. 2018.
21. J. M. González-González, A. Triviño-Cabrera, and J. A. Aguado, "Model predictive control to maximize the efficiency in EV wireless chargers," *IEEE Trans. Ind. Electron.*, vol. 69, no. 2, pp. 1244–1253, Feb. 2022.
22. Z. Zhou, L. Zhang, Z. Liu, et al., "Model predictive control for the receiving-side DC–DC converter of dynamic wireless power transfer," *IEEE Trans. Power Electron.*, vol. 35, no. 9, pp. 8985–8997, Sept. 2020.
23. C. Qi, Z. Lang, L. Su, et al., "Finite-control-set model predictive control for a wireless power transfer system," in 2019 IEEE Int. Symp. Predic. Control Electrical Drives Power *Electron (PRECEDE)*, Quanzhou, China, 2019, pp. 1–5.
24. S. Chen and W. Ding, "A novel finite-control-set model predictive voltage control for dynamic improvement of wireless power transfer systems," in *2021 IEEE 4th International Electrical and Energy Conference (CIEEC)*, Wuhan, China, 2021, pp. 1–6.
25. Y. Yang, "A passive augmented circuit for EMI reductions of full-bridge inverters with conventional phase shift control in wireless power transfer systems," *IEEE Trans. Power Electron.*, vol. 38, no. 11, pp. 13286–13297, Nov. 2023.
26. K. Chen, K. W. E. Cheng, Y. Yang, and J. Pan, "A fast self-positioning-based optimal frequency control for inductive wireless power transfer systems without communication," *IEEE Trans. Ind. Electron.*, vol. 70, no. 1, pp. 334–343, Jan. 2023.
27. T. Tan, K. Chen, Q. Lin, et al., "Impedance shaping control strategy for wireless power transfer system based on dynamic small-signal analysis," *IEEE Trans. Circuits Syst. I: Regul. Pap.*, vol. 68, no. 3, pp. 1354–1365, Mar. 2021.
28. Y. Yang, S. -C. Tan, and S. Y. R. Hui, "Adaptive reference model predictive control with improved performance for voltage-source inverters," *IEEE Trans. Control Syst. Technol.*, vol. 26, no. 2, pp. 724–731, Mar. 2018.

6

Application Domains for Transmitter-Side Control Schemes

6.1 Consumer Electronics and Medium-Power Devices

The "Qi2" standard, introduced by the Wireless Power Consortium (WPC), addresses the growing demand for higher power and higher frequency operations in consumer electronics and emerging medium-power applications such as robotics and drones [1]. This standard emphasizes maximum efficiency transfer (MET) and minimum charging time (MCT) as its core performance metrics. The transmitter-side control strategies presented in this book, which are designed for WPT systems with simplified receiver structures, align well with the principles of the "Qi2" standard. In fact, many of these control schemes can be either directly applied or readily adapted to simultaneously achieve both MET and MCT.

While transmitter-side control centralizes all regulation burdens on the inverter, maintaining soft-switching operation presents significant challenges, particularly for phase shift control schemes across wide operational ranges. Although phase shift control offers the advantage of fixed-frequency operation (a feature incorporated in numerous Qi-standard compliant designs [2]), its implementation difficulties have motivated the proposal of auxiliary passive circuits [3]. I believe such circuits can be adapted into various topologies to enhance flexibility. Nevertheless, as WPT systems scale up in frequency and power, component selection becomes increasingly challenging, introducing new design constraints and performance trade-offs.

6.2 Electric Vehicles and Other Automotive Systems

Dynamic WPT for EVs and other automotive systems, such as automated guided vehicles (AGVs), is a highly promising charging approach that can reduce battery capacity requirements while improving overall power utilization. A photograph of a dynamic WPT laboratory setup is shown in Figure 6.1. In dynamic WPT applications, MCT becomes a critical performance metric, as the charging infrastructure is typically costly and

DOI: 10.1201/9781003468523-6

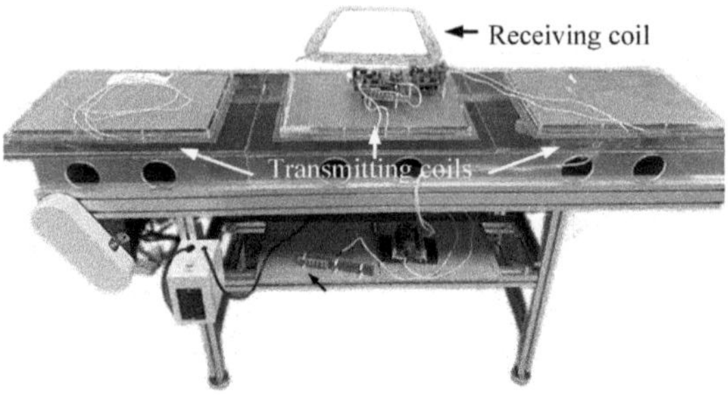

FIGURE 6.1
A photograph of a dynamic WPT laboratory setup.

demands high power delivery. Consequently, even a modest 5% energy loss due to additional receiver-side circuitry can become significant and must be minimized. The transmitter-side control paradigm introduced in this book helps to address this issue by eliminating the need for complex receiver-side circuitry. Furthermore, conventional out-of-band communication systems can introduce delays in high-speed receivers like EVs, particularly during dynamic charging. The transmitter-side control schemes presented here operate independently of feedback from the receiver, effectively mitigating such delays. Finally, automotive manufacturers tend to favor simpler and more reliable receiver designs. By shifting control complexity to the transmitter side, manufacturers of WPT transmitters assume more of the system's risk, naturally driving them to enhance the reliability of their products.

6.3 Underwater Wireless Charging Devices

Autonomous underwater vehicles (AUVs) offer distinct advantages over free-drifting floats and underwater gliders in terms of maneuverability and mission precision, making them indispensable for seabed mapping, deep-sea mineral extraction, and subsea infrastructure inspection [4]. However, their operational endurance remains severely constrained by onboard energy storage limitations, with conventional battery systems typically supporting less than 24 hours of continuous operation. While battery swapping provides a straightforward recharging solution, this approach introduces substantial mission downtime (requiring vehicle recovery, manual battery replacement, and redeployment) and necessitates crew involvement, significantly reducing operational efficiency.

To overcome these limitations, physical docking stations equipped with either plug-in or WPT capabilities have been developed. Traditional plug-in underwater charging typically relies on wet-mate connectors, which require high-precision docking and are prone to safety and maintenance issues due to biofouling and corrosion of the contact surfaces. In contrast, WPT technology provides a more reliable and automation-friendly alternative by eliminating the need for direct electrical contact. The potential of WPT in AUV applications was acknowledged by the U.S. Department of Energy in 2019, recognizing its role in improving charging reliability and operational autonomy. At the heart of any WPT system is the magnetic coupler, which governs the magnetic field distribution and plays a critical role in determining performance metrics such as power transfer efficiency, load power delivery, and output stability.

However, small-scale operations near the seabed remain a significant challenge for existing AUVs with torpedo-like, water-droplet, or elongated cylindrical designs. These traditional forms suffer from limited maneuverability in the horizontal plane and poor vertical stability [5]. To address this issue, a disc-shaped Autonomous Underwater Helicopter (AUH) was recently developed. The AUH is capable of operating directly on the seafloor, and can take off, land, and hover at variable altitudes [6]. This enhanced functionality enables scientific exploration and monitoring in complex underwater environments, such as coral reefs and hydrothermal vents.

One of the major challenges faced by the AUH is its limited endurance due to constrained battery capacity. The current approach involves underwater recharging via a WPT system, eliminating the need for the AUH to surface. However, the existing wireless charging method relies on time-consuming docking procedures prior to power transfer, which increases the risk of collisions and raises maintenance costs. On average, the docking process takes around 20 minutes, yet the actual charging only requires about 8 minutes, making docking more than twice as time-consuming as charging. Furthermore, the precision positioning system required for close-range navigation during docking is costly. This highlights the fact that out-of-band communication between the WPT system's transmitter and receiver is not only unreliable but also financially burdensome. The transmitter-side control schemes presented in this book offer significant potential to overcome communication challenges. The photograph of an underwater WPT system designed for the AUH is shown in Figure 6.2.

FIGURE 6.2
The photograph of an underwater WPT system designed for the AUH.

6.4 Implantable Medical Devices

Driven by an aging population, increasing health awareness, and rapid technological advancements, the global market for implantable medical devices (IMDs) is experiencing robust growth [7]. In 2024, the market was valued at approximately $131.9 billion and is projected to reach around $306.8 billion by 2034, reflecting a compound annual growth rate (CAGR) of 8.9% from 2025 to 2034 [8]. Among IMDs, devices that operate without requiring precise positioning (such as wireless capsule endoscopes (WCEs) that traverse the esophagus, stomach, and small intestine) represent significant market potential. The global market for WCEs is notably strong. According to a 2019 report by the International Foundation for Gastrointestinal Disorders, gastroesophageal reflux disease (GERD) affects more than one-third of the global population, leading to bleeding complications that fuel the demand for capsule endoscopy. The global capsule endoscopy market was valued at approximately USD 682.6 million in 2022 and is expected to grow at a CAGR of 9.5%, reaching nearly USD 1.6 billion by 2032 [9]. In China, the capsule endoscopy sector has also demonstrated consistent growth. The market size expanded from 380 million CNY in 2018 to 650 million CNY in 2022, indicating a CAGR of 14.2%. It is projected to reach 4.32 billion CNY by 2027, with a CAGR of 45.9% from 2022 to 2027 [10].

However, IMDs face significant limitations due to their reliance on onboard batteries [11]. These power sources impose two critical constraints: (i) finite operational lifespans necessitating invasive replacement surgeries, and (ii) physical size requirements that directly impact device miniaturization. WPT technology presents a transformative solution by enabling both battery lifespan extension through opportunistic charging and substantial reduction in required battery capacity via continuous power supplementation. Currently, companies developing WPT technologies for charging IMDs are still in the early stages of growth. Resonant Link, founded in 2017 in the United States, offers wireless charging solutions for neurostimulators, infusion pumps, and pacemakers, utilizing its proprietary Multi-Layer Self-Resonant Structure (MSRS) technology. Powermat, an Israeli company established in 2006, entered the IMD market in 2022 with offerings such as the PMT 100, a 40 W wireless power solution specifically designed for medical applications. WPT systems IMDs demand compact designs and highly reliable operation. Minimal human intervention in maintaining the receiver is always preferred. In emerging applications (such as artificial hearts and microrobots for treating Parkinson's disease) receiver units must be fully enclosed and based on extremely simple circuit topologies. As such, transmitter-side control schemes that operate without external communication feedback are particularly well-suited for these scenarios. Figure 6.3a presents a schematic diagram of the WPT system designed for a capsule endoscope. Figure 6.3b displays the laboratory-scale implementation of the WPT transmitter.

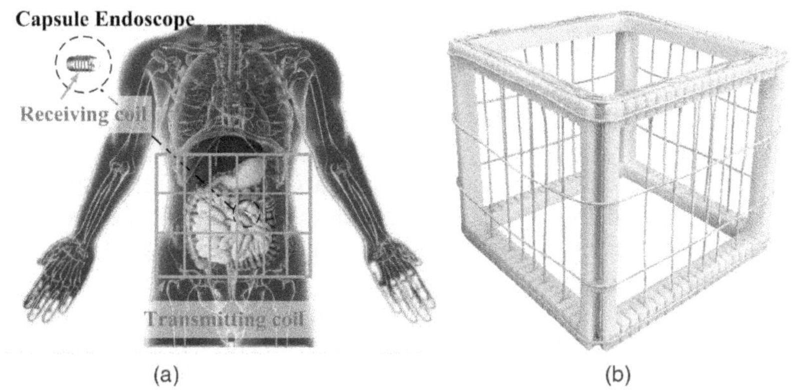

FIGURE 6.3
(a) Schematic diagram of the WPT system designed for a capsule endoscope, and (b) its physical realization of the transmitter.

6.5 Future Research Perspectives

With advancements in semiconductor technologies, WPT systems are expected to operate at higher power levels and frequencies, provided that electromagnetic compatibility (EMC) issues are effectively managed. This progress will enable greater spatial freedom in WPT applications. Concepts currently at the laboratory stage, such as wireless charging rooms, may soon become practical. The transmitter-side monitoring strategies presented in this book could be applied to detect the positions of power receivers (even for movable devices like drones) within a wireless charging chamber, without relying on out-of-band communication. Additionally, the proposed transmitter-side control schemes can ensure safe charging operations even when the chamber functions as a communication-isolated environment, such as a Faraday cage. However, I view this as a natural extension of the current research, focused more on new applications of the existing transmitter-side monitoring and control framework rather than a fundamental shift in the approach. I have the same perspective regarding the use of quantum communication, which, in the context of WPT applications, could essentially be viewed as "no communication". If quantum communication becomes widely commercialized in the future, transmitter-side control schemes could be significantly simplified.

I think the true breakthrough in adopting the new control paradigm by shifting all control algorithms to the transmitter side relies heavily on advancements in AI technologies. As AI becomes more deeply integrated into the manufacturing processes of future WPT systems, which will operate at higher power levels and frequencies, an increasing amount of component-level data (such as aging behavior, thermal characteristics, and structural deformation) will be collected and stored in the cloud. These data will contribute to the creation of digital twins for each component, which can be virtually assembled into complete digital WPT systems. Measurement data from physical systems around the world can then be integrated into these digital counterparts, helping to bridge the gap between digital and physical performance. This growing digital ecosystem will enable the formation of extensive datasets that can be used to train highly sophisticated control models (including transmitter-side controllers) capable of achieving near-perfect performance: zero steady-state error, optimal dynamic response, zero overshoot, minimal harmonics, and more. All of the AI-based control concepts discussed in previous chapters could be realized within this framework. That said, I also see great potential in the development of AI-based controllers that require minimal or even no explicit input parameters, instead generating control variables (such as phase shift angle or switching frequency) through learned time-sequence mappings. Of course, this is just my perspective. AI is evolving rapidly and unpredictably, often in ways that surpass our current imagination.

References

1. Qi v2.0 [Online]: https://www.chargewithqi.com/media/w0ha5cbk/qi2-certification-rolls-out-news-release-11132023.pdf.
2. QiSpecification[Online]:https://www.wirelesspowerconsortium.com/knowledge-base/specifications/download-the-qi-specifications/.
3. Y. Yang, "A passive augmented circuit for EMI reductions of full-bridge inverters with conventional phase shift control in wireless power transfer systems," *IEEE Trans. Power Electron.*, vol. 38, no. 11, pp. 13286–13297, Nov. 2023.
4. L. Paull, S. Saeedi, M. Seto, and H. Li, "AUV navigation and localization: a review," *IEEE J. Ocean. Eng.*, vol. 39, no. 1, pp. 131–149, Jan. 2014.
5. Z. Wang, L. Xun, H. Huang, and Y. Chen, "Development of an autonomous underwater helicopter with high maneuverability," *Applied Sciences*, vol. 9, no. 19, pp. 4072, Sept. 2019.
6. J. Zhou, H. Huang, S. H. Huang, et al., "AUH, a new technology for ocean exploration," *Engineering*, vol. 25, pp. 21–27, Jun. 2023.
7. K. Agarwal, R. Jegadeesan, Y. -X. Guo and N. V. Thakor, "Wireless power transfer strategies for implantable bioelectronics," *IEEE Rev. Biomed. Eng.*, vol. 10, pp. 136–161, 2017.
8. Global Market Insights, "Implantable medical devices market size, 2025 to 2034," GMI4483, Feb 2025 [Online]: https://www.gminsights.com/industry-analysis/implantable-medical-devices-market
9. Global Market Insights, "Capsule endoscopy market size by product, application, and end-use – global forecast, 2023 to 2032," GMI5928, Jun. 2023 [Online]: https://www.gminsights.com/industry-analysis/capsule-endoscopy-market

10. Chinabgao, "2024 Capsule endoscopy market analysis: market size exceeds 650 million CNY," Dec. 23, 2024 [Online]: https://www.chinabaogao.com/free-report/97771.html
11. S. Yang, V. Sencadas, S. S. You, N. Z.-X. Jia, S. S. Srinivasan, H. Huang, J. Y. Liang, and G. Traverso, "Powering implantable and ingestible electronics," *Adv. Funct. Mater.*, vol. 31, no. 6, p. 2009289, Feb. 2021.

Index

For Product Safety Concerns and Information please contact our EU
representative GPSR@taylorandfrancis.com
Taylor & Francis Verlag GmbH, Kaufingerstraße 24, 80331 München, Germany

www.ingramcontent.com/pod-product-compliance
Ingram Content Group UK Ltd.
Pitfield, Milton Keynes, MK11 3LW, UK
UKHW020206200926
475702UK00004B/640
* 9 7 8 1 0 3 2 7 3 7 8 1 2 *